Asphaltene Particles in Fossil Fuel Exploration, Recovery, Refining, and Production Processes

Asphaltene Particles in Fossil Fuel Exploration, Recovery, Refining, and Production Processes

Edited by

Mahendra K. Sharma
Eastman Chemical Company
Kingsport, Tennessee

and

Teh Fu Yen
University of Southern California
Los Angeles, California

Springer Science+Business Media, LLC

Library of Congress Cataloging in Publication Data

Asphaltene particles in fossil fuel exploration, recovery, refining, and production processes / edited by Mahendra K. Sharma and Teh Fu Yen.

p.   cm.

"Proceedings of an International Symposium on Asphaltene Particles in Fossil Fuel, Exploration, Recovery, Refining, and Production Processes, held in conjunction with the Twenty-third Annual Meeting of the Fine Particle Society, July 13–17, 1992, in Las Vegas, Nevada"—Galley.

Includes bibliographical references and index.

ISBN 978-0-306-44709-9   ISBN 978-1-4615-2456-4 (eBook)

DOI 10.1007/978-1-4615-2456-4

1. Asphaltene—Congresses. I. Sharma, Mahendra K. II. Yen, Teh Fu, [DATE]. III. International Symposium on Asphaltene Particles in Fossil Fuel Exploration, Recovery, Refining, and Production Processes (1992: Las Vegas, Nev.) IV. Fine Particle Society, Meeting (23rd: 1992: Las Vegas, Nev.)

TP692.4.A8A73 1994	94-11708
665.5'388—dc20	CIP

Proceedings of an International Symposium on Asphaltene Particles in Fossil Fuel Exploration, Recovery, Refining, and Production Processes, held in conjunction with the Twenty-Third Annual Meeting of the Fine Particle Society, July 13–17, 1992, in Las Vegas, Nevada

ISBN 978-0-306-44709-9

© 1994 Springer Science+Business Media New York
Originally published by Plenum Press, New York in 1994

PREFACE

THE CURRENT STATE OF THE ART of several aspects of asphaltene is presented in this volume. It documents the proceedings of the Internationl Symposium on Asphaltene Particles in Fossil Fuel Exploration, Recovery, Refining and Production Processes sponsored by the Fine Particle Society (FPS). This meeting was held in Las Vegas, Nevada, July 13-17, 1992. The symposium upon which this volume is based was organized in four sessions emphasizing various basic and applied aspects of research on asphaltene technology. Major topics discussed involve surface phenomena of asphaltene, processed and unprocessed bitumen, asphaltene effect on natural and accelerated ageing of bitumens, asphaltene conversion, theoretical aspects of asphaltenes and interactions of asphaltene colloids in organic solvents.

This edition includes eighteen selected papers presented at the symposium. These papers are divided in four broad categories: (1) Bitumen and Coal-Derived Asphaltenes, (2) Asphalt and Asphaltene Conversion, (3) Surface and Colloidal Aspects of Asphaltenes, and (4) Thermodynamic and Molecular Aspects of Asphaltenes.

This proceedings volume includes discussions of various processes occuring at molecular, microscopic, and macroscopic levels in asphaltenes and bitumen processing for heavy oil recovery. The editors hope that this volume will serve its intended objective of reflecting the current understanding of formulation and process problems related to asphaltenes. In addition, it will be a valuable reference source for both novices as well as experts in the field of asphaltene and petroleum technology. It will also help the readers to understand underlying surface phenomena and will enhance the reader's potential for solving critical formulation and process problems.

The editors would like to convey their sincere thanks and appreciation to the Fine Particle Society for the generous support. We would also like to express our thanks and appreciation to Ms. Patricia M. Vann and to the Editorial Staff of the Plenum Publishing Corporation for their continued interest in this project.

The editors are grateful to reviewers for their time and efforts in providing valuable comments and suggestions to improve the material presented in the manuscripts. We wish to convey our sincere thanks and appreciation to all authors and coauthors for their contributions, enthusiasm and patience. The views and conclusions expressed herein are those of the authors.

One of us (MKS) would like to express his thanks to the appropriate management of the Eastman Chemical Company (ECC) for allowing him to participate in the organization of the symposium. His special thanks are due to Mr. J. C. Martin (ECC) for his cooperation and understanding during the tenure of editing this proceedings volume.

Finally, MKS wishes to express his sincere thanks to his colleagues and friends for their assistance and encouragement throughout this project. Also he would like to acknowledge the assistance and cooperation of his wife, Rama, and extends his appreciation to his children (Amol and Anuj) for allowing him to spend many evenings and weekends working on this volume.

M. K. Sharma
Research Laboratories
Eastman Chemical Company
Kingsport, TN 37662

T. F. Yen
Environmental and Civil Engineering
University of Southern California
Los Angeles, CA 90089

CONTENTS

THERMODYNAMIC AND MOLECULAR ASPECTS OF ASPHALTENES

ASPHALTENE - VISCOSITY RELATIONSHIP OF PROCESSED AND UNPROCESSED BITUMEN

A. Chakma[*], M.R. Islam[+] and F. Berruti

Dept. of Chemical and Petroleum Engineering
University of Calgary
Calgary, Alberta, CANADA, T2N 1N4

[+]Dept. of Geological Engineering
South Dakota School of Mines
Rapid City, SD 57701, USA

ABSTRACT

Bitumen and heavy oils produced in Alberta frequently contain asphaltenes. These asphaltenes may play a significant role in the primary upgrading or visbreaking of the bitumen and heavy oils. This paper describes an experimental study of the effect of asphaltene content on the viscosity of bitumen. Rheological studies were carried out using virgin bitumen as well as using mixtures of de-asphalted bitumen and asphaltenes. The viscosity of the mixture was found to be a strong function of asphaltene content. However, it was not possible to ascertain the exact nature of this functionality. In order to obtain further insight into this, processed samples were also included in the study. Since, the aggregate nature of the asphaltene is subject to change when subjected to the high severity conditions of upgrading reactors, the use of processed sample was expected to provide some additional information on the asphaltene content-viscosity relationship. It was found that asphaltenes with higher molecular weight resulted in higher viscosity of the mixture than those with lower molecular weights.

---

[*]Author to whom correspondence should be addressed

Asphaltene Particles in Fossil Fuel Exploration, Recovery, Refining, and Production Processes, Edited by M.K. Sharma and T.F. Yen, Plenum Press, New York, 1994

1

INTRODUCTION

In many petroleum reservoirs around the world, reservoir fluid composition has been found to vary with location and depth.[1-3] In almost all of the cases, reservoir fluid density increases with the depth of the reservoir[1]. Tar sand bitumen deposits found in the Province of Alberta are no exceptions. Patel[4] found the viscosity of Athabasca, Peace River, Wabasca and Cold Lake bitumen vary with the depth of the formation. In most cases, samples obtained from higher depths showed greater viscosity. Schutze[3] studied the compositional variations within a hydrocarbon column due to gravity and found the gravitational forces to be responsible for the variation in composition in thick reservoirs. He found the extent of this variation to be higher with larger aromatic fractions in the hydrocarbon fluid.

Hirschberg[5], in his analysis of the role of asphaltenes in the compositional variation of a reservoir fluid column concluded that the heavy polar compounds play a key role in this regard. He found asphaltene segregation to have a dominant effect in the process. One of the principal effect of compositional changes is the variation in viscosity. For a reservoir oil sample from a North African field, Hirschberg[5] found the viscosity to increase by a factor of 4 (from 9 to 36 mPa.s) when the asphalt content increased from 10 to 16%. Indeed, the effect of asphaltene concentration on the viscosity of oil has been known for a long time. Mack[6], in 1932, presented viscosity data on asphaltene and oil mixtures obtained from a Mexican asphalt that clearly showed an increase in viscosity with increasing asphaltene concentration. Waxman et al.[7] as well as Kitzan and Parsons[8] have found the viscosity of Peace River bitumen samples to vary with asphaltene concentration. Dealy[9] studied the effect of asphaltene concentration on the viscosity of Athabasca bitumen by adding 5 wt% of additional asphaltene to a bitumen sample that originally contained 16 wt% of asphaltene. The viscosity of the bitumen was found to increase from about 300 Pa.s to 1000 Pa.s as a result of asphaltene addition. When higher shear was applied, the viscosity of the mixture was found to decrease somewhat, but was still above that of the original bitumen.

Altgelt and Harle[10] also studied the effect of asphaltenes on asphalt viscosity. They found the asphaltenes to form aggregates in solution, the degree of which was found to depend on the structure, molecular weight and concentration of the asphaltenes as well as the power of the solvent. They concluded that the viscosity of asphaltene containing fluids is primarily due to the aggregation of the asphaltenes.

Chakma and Berruti[11] studied the effect of ultrasound on the viscosity of Athabasca bitumen and found that the viscosity reduction of up to 15% can be achieved by the application of ultrasound. They concluded that ultrasound causes the asphaltene aggregates to break up to some extent.

As a result, viscosity reduction occurs. Therefore, there is ample evidence that clearly shows that the viscosity of the bitumen is a strong function of asphaltene concentration and nature.

Physicochemical Background

Alberta bitumen may be considered to be composed of macro- and micro- structures as well as chemical constitutive molecules. The macro and micro structural arrangements determine the viscosity of the bitumen. Dace and Yen[12] investigated the macro-structures of the asphaltenes and found them to be composed of polynuclear aromatic molecules with alkyl chains as attachments. These constitutive "asphaltene unit molecules" are grouped in layers having several unit molecules (typically 5 or 6) surrounded or immersed into the maltese (fraction soluble in 40 volume of n-pentane) fluid. The latter is composed of free saturates, mono- and d-aromatics and resins that may be associated with the asphaltenes. This structural organization may be considered to be the microstructure of the system. The microstructure may form aggregates in order to reduce the free energy of the system. These aggregates form micellar structures which consist of several unit layers of asphaltenes surrounded by, or associated with the maltenes. The micellar structures thus formed may be classified as the macro-structures of the system.

The macro-structural arrangement to a large extent depends on the size of the unit asphaltene layers. Low molecular weight asphaltenes (Mw up to about 1000) consist of single sheets of condensed aromatic and naphthenic rings with relatively short alkyl chain attachments. Larger molecular weight asphaltenes consist of several sheets of condensed ring systems connected to each other by short chains[12,13]. Altgelt and Harle[11] found that the aggregates formed by low molecular weight asphaltenes (Mw less than 10,000) result in only a slight increase in viscosity due to their relatively compact structures. However, those formed by larger molecular weight asphaltenes may increase viscosity substantially.

The molecular weights of petroleum asphaltenes are very high and range from 1,000 to 2,000,000. This wide variation in asphaltenes' molecular weight is partly due to the different measuring techniques used in its determination but mostly due to the association of the asphaltene units at the condition of measurements. For example, highly associated asphaltenes will give higher molecular weight than the mildly associated ones. Among the various techniques available, vapor pressure osmometry is the most frequently used one. Ultracentrifuge and electron microscopy based techniques give very high molecular weights. Molecular weight determination by vapor pressure osmometry may not always provide consistent results as the molecular weight is affected by the nature of the solvent and its dielectric constant.

On the basis of the above discussion, any attempt to quantify the effect of asphaltenes on the viscosity of bitumen must include molecular weight and concentration of the asphaltenes along with the characteristics of the solvent.

In light of the above, the present work was undertaken to elucidate the relationship between asphaltene concentration and viscosity of bitumen by analyzing existing literature data and by conducting rheological studies in the laboratory.

EXPERIMENTAL

The experimental work essentially consisted of rheological measurements of Athabasca bitumen, de-asphaltene Athabasca bitumen, mixtures of virgin Athabasca bitumen and asphaltenes, products obtained from hydrocracking of Athabasca bitumen and products obtained from co-processing of Athabasca bitumen and coal with or without the presence of any catalysts.

Equipments

The equipment used for rheological measurements was a rotational rheometer (Rheomat 115 equipped with a programmable microprocessor based temperature and speed controller, Rheoscan 110, Contraves, Zurich, Switzerland).

Samples

Athabasca bitumen samples were obtained from the Alberta Research Council Sample Bank. It contained 15.4 % asphaltene.

Asphaltene samples were separated from the virgin Athabasca bitumen by first dissolving the bitumen in toluene. The toluene insoluble were separated and the toluene was then evaporated in a Rotavapor (Bucchi, Switzerland). Asphaltenes were then precipitated by adding 40 volumes of n-pentane.

Hydrocraked Athabasca bitumen samples were prepared in our laboratory by reacting Athabasca bitumen under hydrogen pressure at different temperatures and reaction times. Similarly, co-processed samples were also obtained from our laboratory.

RESULTS AND DISCUSSION

Athabasca Bitumen

Figure 1 shows the viscosity of virgin and de-asphalted Athabasca bitumen as a function of temperature. It is clear that with asphaltene removal, a rapid reduction in viscosity

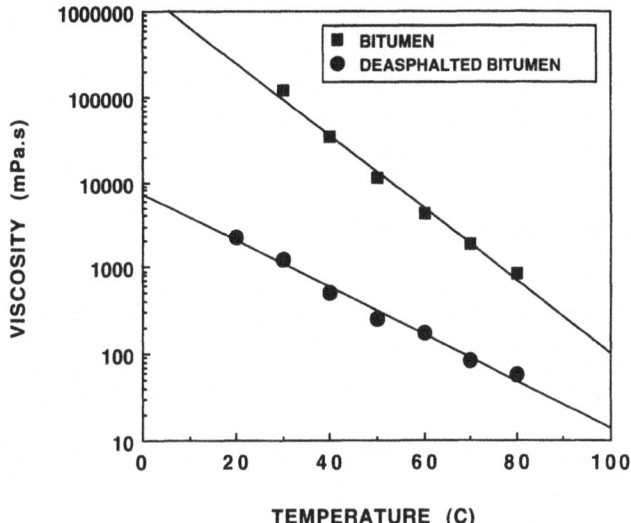

Figure 1. Viscosity of virgin Athabasca bitumen (15.4%
asphaltene) and de-asphalted Athabasca bitumen as
a function of temperature.

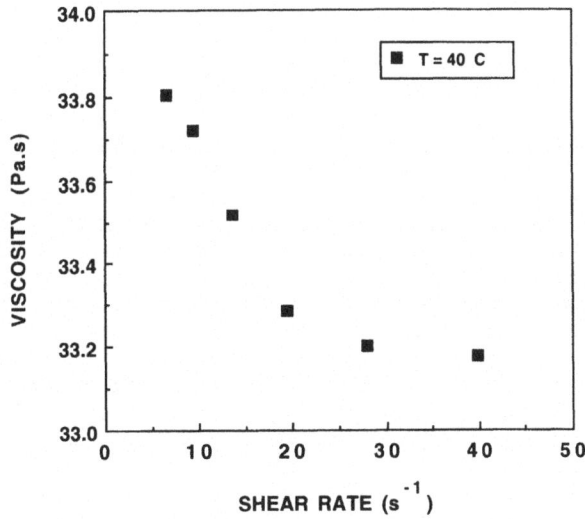

Figure 2. Viscosity of Athabasca bitumen as a function of
shear rate.

is achieved. The effect of temperature on the viscosity of the two samples is also noteworthy. While, viscosity decreases rapidly with temperature for the two samples, the effect is more pronounced for the virgin bitumen compared to the de-asphalted bitumen as can be seen from the slopes of the two curves. As the temperature rises, the asphaltene aggregates in the virgin bitumen gain more fluidity due perhaps to some degree of desegregation.

Figure 2 shows viscosity of virgin Athabasca bitumen measured at 40 °C, as a function of shear rate. The viscosity is clearly affected by the shear rate, but the change is not significant. However, it can be said that the virgin bitumen is slightly non-Newtonian.

When 5 wt% of asphaltene was added to the virgin Athabasca bitumen at 40 °C, its viscosity increased in excess of two folds. Figure 3 shows the viscosity of virgin sample and that of the mixture as a function of shear rate. It was difficult to mix 5 wt% of asphaltenes with the virgin bitumen at room temperature. Hence, the measurements were made at 40 °C. Due to the change in scales, the mild non-Newtonian behavior of the virgin bitumen can no longer be observed from this figure. However, the mixture containing additional asphaltene clearly shows a non-Newtonian behavior. This figure indicates that with high shear, viscosity of asphaltene containing mixture decreases. This can be attributed to the desegregation of the asphaltene aggregates.

Processed Bitumen

In the case of unprocessed bitumen, it was possible to obtain some qualitative idea about the effect of asphaltenes on viscosity and rheological behavior of the bitumen. However, the quality, (i.e. composition, degree of association, etc.) of the asphaltene in the virgin bitumen and that of the separated bitumen may not have been the same due to the effects of the solvents used during the separation process. Since asphaltenes do not occur as "pure" components in the bitumen formations, it was not possible to obtain them in such state. When bitumen is processed in operations such as hydrocracking, thermal visbreaking, bitumen-coal co-processing, etc., not only the concentration of the asphaltenes change, the nature of the asphaltenes changes as well due to the cracking of the molecules. As such, products from bitumen upgrading operations were considered to be a good source of asphaltenes having different concentrations and qualities. Since, molecular weight of the asphaltenes is a good indication of their degree of association, average molecular weights in their maltese solvents were also measured by vapor pressure osmometry.

Figure 4 shows viscosity as a function of asphaltene concentration in products obtained from hydrocracking of Athabasca bitumen in autoclaves under hydrogen pressure at

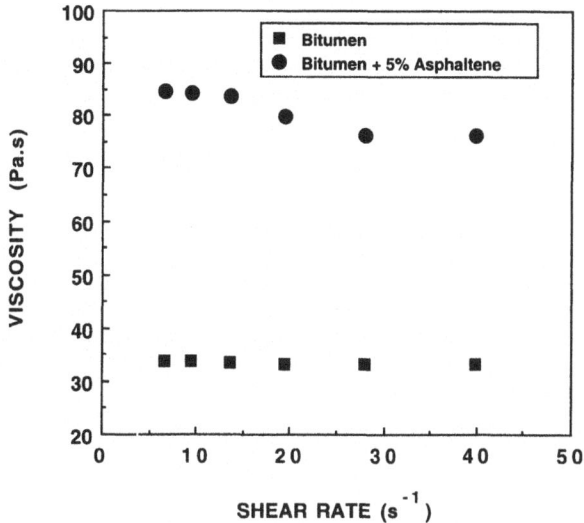

Figure 3. Viscosity of Athabasca bitumen and a mixture of Athabasca bitumen and 5 wt% asphaltene as a function of shear rate (Temp. 40 °C).

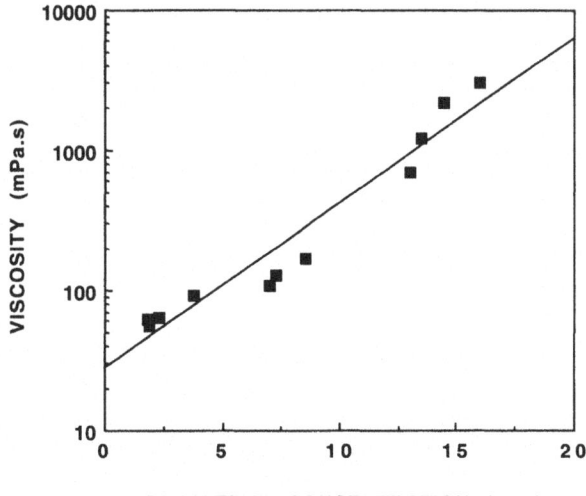

Figure 4. Viscosity of products obtained by hydrocracking of Athabasca bitumen as a function of asphaltene concentration (Temp: 25°C).

different temperature and reaction time conditions. A clear relationship between viscosity and asphaltene concentration in the product can be seen from Figure 4.

Figure 5 shows viscosity as a function of average molecular weight of the hydrocracked products. Again, there seems to be a relationship between viscosity and molecular weight of the cracked products.

Bitumen-Coal Coprocessing Products

Figure 6 shows the viscosity of liquid products obtained by coprocessing of Athabasca bitumen with a sub-bituminous coal in the presence of hydrogen. Figure 7 shows the viscosity as a function of average molecular weight of the liquid products. As with the previous cases, a distinct relationship between viscosity, asphaltene concentration and molecular weight is evident.

Jobo Heavy Oil

Along with our own experimental results, here we include the data from Schutze[14] on Jobo heavy oil that essentially shows the same trends.

Figure 8 shows the viscosity of the liquid products obtained by Schutze[14] by hydrocracking Jobo heavy oil (initially containing 16.25% of asphaltenes), carried out in a Spray reactor, as a function of asphaltene concentrations. Figure 9 shows viscosity as a function of average molecular weight for the same products. As with Athabasca bitumen, the products of the Jobo heavy oil also tends to show relationships between viscosity, asphaltene concentration and average molecular weight. However, in this case there is considerable scatter in the data.

CONCLUSIONS

The effect of asphaltene on the viscosity of unprocessed and processed bitumen has been investigated. While the virgin bitumen was found to be mildly non-Newtonian, addition of asphaltene adds to the deviation from the Newtonian behavior. The results presented in this paper clearly shows that viscosity of bitumen and heavy oil is a function of asphaltene concentration, molecular size, and the properties of the solvent. Average molecular weight of the mixture is a parameter, which provides a lumped representation of the molecular weight of the asphaltenes and the properties of the solvent. Hence, it is possible to relate viscosity of bitumen to asphaltene concentration and the average molecular weight of the mixture.

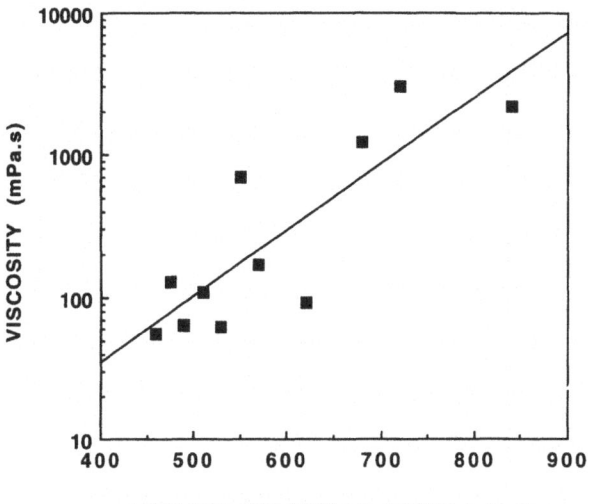

Figure 5. Viscosity of products obtained by hydrocracking
of Athabasca bitumen as a function of average
molecular weight (Temp: 25°C).

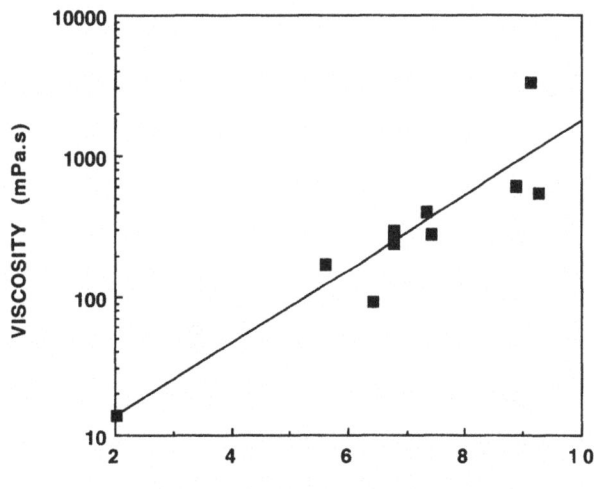

Figure 6. Viscosity of liquid products obtained by
coprocessing Athabasca bitumen and Coal, as a
function of asphaltene concentration
(Temp: 25 °C).

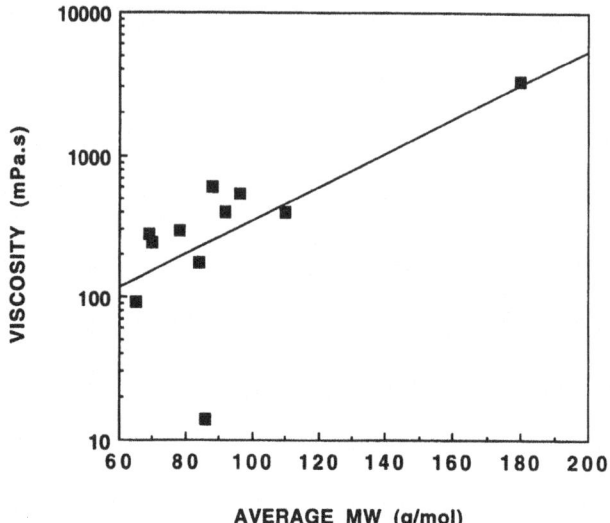

Figure 7. Viscosity of liquid products obtained by coprocessing Athabasca bitumen and Coal as a function of average molecular weight (Temp: 25 °C).

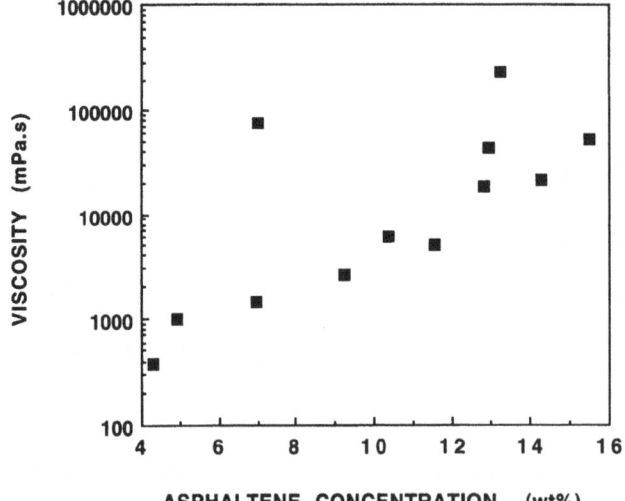

Figure 8. Viscosity of liquid products obtained by cracking Jobo heavy oil as a function of asphaltene concentration (Temp. = 25 °C, Data: Schutze[14]).

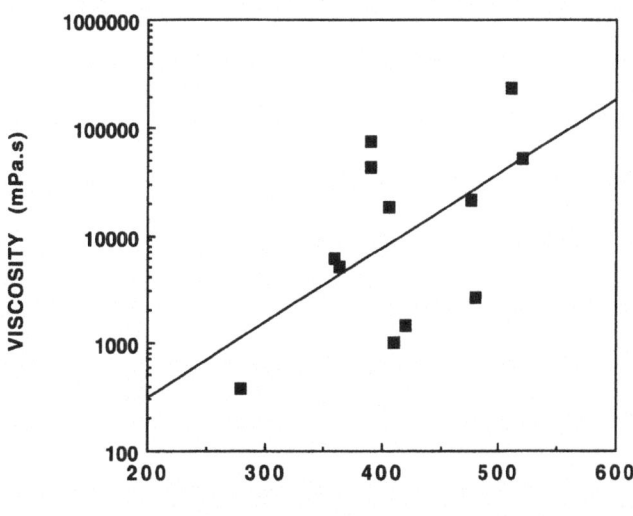

Figure 9. Viscosity of liquid products obtained by cracking
Jobo heavy oil as a function of asphaltene
concentration (Temp. = 25 °C, Data: Schutze[14]).

ACKNOWLEDGMENTS

Financial support of the Natural Sciences and Engineering Research Council of Canada is gratefully acknowledged. Mr. Yuenfeng Hu and Mr. C.M. Talukder assisted in the experimental work.

REFERENCES

1. Sage, B.H., Lacey, W.N.: "Gravitational Concentration Gradients in Static Columns of Hydrocarbon Fluids", Trans. AIME, 132, 122-131, 1939.

2. Hunt, M.: "Petroleum Geochemistry and Geology", W.H. Freeman and Co., San Francisco, pp 381-391, 1979.

3. Schulte, A.M.: "Compositional Variations Within a Hydrocarbon Column Due to Gravity", SPE 9235, 1980.

4. Patel, M.S.: "Determination of Viscosities of Oils From Mannville Formation Oil Sand", Alberta Research Council Report, October, 1973.

5. Hirschberg, A.: "The Role of Asphaltenes in Compositional Grading of a Reservoir's Fluid Column", SPE 13171, 1984.

6. Mack, C.: "Colloid Chemistry of Asphalts", J. of Phys. Chem., 36, 2901, 1932.

7. Waxman, M.H., Deeds, C.T., Closmann, P.J.: "Thermal Alteration of Asphaltenes in Peace River Tars", SPE 9510, 1980.

8. Kitzan, P., Parsons, L.J.: Quoted in "The Thermodynamic and Transport Properties of Bitumens and Heavy Oils", AOSTRA, Edmonton, AB., 1984.

9. Dealy, J.M.: "Rheological Properties of Oil Sands Bitumens", Can. J. of Chem. Eng., 57, 677-683, 1979.

10. Altgelt, K.H., Harle, O.L.: "The effect of Asphaltenes on Asphalt Viscosity", Ind. Eng. Chem. Prod. Res. Dev. 14(4), 240-246, 1975.

11. Chakma, A., Berruti, F.: "Ultrasonic Visbreaking of Athabasca Bitumen", Proc. of the 5th UNITAR International Conference on Heavy Crude and Tar Sands, Vol. 4, 101-104, 1991.

12. Dace, J.P., Yen, T.F.: Anal. Chem., 39, 1847, 1967.

13. Ferris, S.W., Black, E.P., Clelland, J.B.: Ind. Eng. Chem. Prod. Res. Dev., 6, 127, 1967.

14. Schutze, B., "Untersuchungen zur Hydropyrolyse von Schwerol und Erdolruckstanden", Bundesministerium fur Forschung und Technologie, Bonn, Germany, 1983.

NATURAL AND ACCELERATED AGEING OF BITUMENS:

EFFECTS OF ASPHALTENE

F. S. Choquest and A. F. Verhasselt

Research and Development Department
Belgian Road Research Institute
1200 Brussels, Belgium

ABSTRACT

A Study on the bitumen ageing has been
conducted by monitoring generic composition
such as saturates, cyclics, resins and
asphaltenes as a   of time. Results
have shown that the ageing process is
represented by the following reaction sequence:

CYCLICS ⟶ RESINS ⟶ ASPHALTENES

It has been found that natural ageing caused
the formation and accumulation of resins in
larger (two to three times) portions than
asphaltenes on monitoring the balance of this
reaction and comparing the compositions of aged
and original bitumens. The effect of various
operating conditions like temperature,
atmospheric conditions, time, equipment etc. on
the formation of asphaltenes has also been
studied. Experimental data showed a significant
qualitative and quantitative differences in the
development of asphaltenes at several ageing
temperatures. In accelerated laboratory ageing,
temperatures of $100^{\circ}C$ or above resulted in the
preferential formation of asphaltenes at an
early stage of ageing process. For temperatures
below approximately $90^{\circ}C$, it has been observed
that the laboratory simulations data are
closely related to the changes observed in road
conditions.

*Asphaltene Particles in Fossil Fuel Exploration, Recovery, Refining, and Production
Processes,* Edited by M.K. Sharma and T.F. Yen, Plenum Press, New York, 1994

13

INTRODUCTION

Bitumen ageing in a road in its climatic environment brings about a change in rheological behavior. It has been reported that the bitumen hardens, and it becomes more brittle at low temperatures. These changes in bitumen may cause in distress in, and consequent deterioration of, the road surfacing on ageing.

The variations in binder consistency referred to above are only exterior signs of intrinsic phenomena which occur in the composition and chemical constitution of the bitumen which governs the progress of ageing. The reactions involved depend on an oxidation process, and are more intense as air and moisture have easier access to the binder in the pavement.

Although determining the chemical composition of bitumens in detail is a difficult task because of its complexity, it is possible, at a much more summary level, to define a number of families of so-called "generic" compounds by means of specific solvent separation techniques and high performance liquid chromatography (HPLC).

The extraction of bitumen by hot n-heptane permits a separation into two major families of compounds: the **maltenes** (soluble) and the **asphaltenes A7** (insoluble). Liquid chromatography in a column filled with alumina and silica and elution with different solvents (successively n-heptane, toluene and 80/20 mixture of toluene and methanol) enable the maltenes to be separated into four fractions of compounds: saturates, cyclics and two types of resins (resins 1 and 2).

Several studies conducted at the Belgian Road Research Center (BRRC) and also at the French "Laboratoire Central des Ponts et Chaussees" (LCPC) on bitumens extracted and recovered from road surfacing poor in voids and of different ages (ranging from four to nineteen years) have shown that the progress of ageing in bitumens can be monitored by examining the variations in their generic compositions. From the compositional point of view, bitumen ageing is represented by the following reaction sequence:

Cyclics ⟶ Resins ⟶ Asphaltenes

An attempt was made to characterize the bitumen ageing process by an increase in content of hot n-heptane insoluble compounds or **asphaltenes A7**. The compositional changes as a function of bitumen ageing were studied by ageing material under accelerated conditions in the laboratory as well as by collecting aged material from the roads (field). The results of field and laboratory investigations were compared in order to examine the effect of several factors on the bitumen ageing process.

NATURAL BITUMEN AGEING IN ROADS

Two independent studies were conducted on the field ageing of bitumens at BRRC and LCPC (Paris). At BRRC, experiments were performed on fifteen road sections of different ages ranging from four to nineteen years, while at LCPC, three different bitumens used in motoway A08 with ageing period of five years were investigated. The results of these studies provide an insight into the role and influence of natural ageing on the generic composition of bituminous binders and in particular on the formation of asphaltenes.

THE BRRC EXPERIMENTS

The following procedure[1] was used to collect, prepare, extract and analyze the bitumen composition:

Collection and Preparation of Samples: Several samples were collected from the road at fifteen different locations as recorded in Table I. Among these locations, thirteen had closed bituminous concrete surfacing with less than 5% of voids and the remaining two had semi-closed surfacing with 6-7% of voids (road N 211 and N 972). All the bitumens, except three bitumens of road N 211, were of 50/70 penetration grade at the time of construction.

Twenty six core samples were collected from each location. The wearing course (5 cm thick) of each sample was separated by sawing with a diamond disc cutter. The top slice of that wearing course (TS = 0.5 cm) was then separated from the "bulk portion" (BP = 4.5 cm).

Extraction and Recovery of the Binders: The binders were extracted separately from the top slice (TS) and the bulk portions (BP) using toluene as a solvent. Each qualitative recovery of bitumen was checked by infrared spectroscopy. For analysis, the absorption band of toluene at the frequency of 700 cm^{-1} had to be absent from the spectrum of the bitumen.

Generic Composition of the Recovered Binders: The experimental data on generic composition of extracted and recovered binders are listed in Table I. In top slices (TS), a strong effect of ageing on several parameters was recorded. An increase in ring ball temperature, a decrease in penetration and an increase in Fraass's breaking point show that the phenomena occurred in the upper few millimeters of the road surface.

Table I shows that the following changes in the generic composition can be recorded when passing from the bulk portion (BP) to the top slices (TP): (1) an increase in asphaltenes content (insoluble in n-heptane) and in resin content; and (2) a decrease in cyclics content. Since the decrease in cyclics content always exceeds the increase in asphaltenes content, it can be assumed that the cyclic

Table I. Generic Composition of Extracted and Recovered Binders.

Road	Sample	Saturat. (S) %	Cyclics (C) %	Resins (R) %	Asphalt. (A₇) %	ΔRes. (R) %	ΔA7 (A₇) %
RO (1969)	T.S.	22.8	33.3	28.0	15.9		
	B.P.	20.5	36.5	28.6	14.4	-0.6	1.5
N 27 (1977)	T.S.	16.3	29.3	31.4	23.0		
	B.P.	20.7	31.6	27.1	20.6	4.3	2.4
N 49 (1977)	T.S.	12.9	38.9	31.4	16.6		
	B.P.	14.8	46.8	26.7	11.7	4.7	4.9
N 53 (1981)	T.S.	15.9	32.0	27.3	24.8		
	B.P.	20.8	38.6	17.7	22.9	9.6	1.9
N 211- T₁ 50/70 (1968)	T.S.	11.1	37.2	37.9	13.8		
	B.P.	11.6	53.8	27.5	7.1	10.4	6.7
N 211- T₂ 50/70 (1968)	T.S.	16.4	31.3	32.1	20.2		
	B.P.	17.5	45.2	21.6	15.7	10.5	4.5
N 211- T₃ 50/70 (1968)	T.S.	15.2	29.4	30.2	25.2		
	B.P.	17.6	40.1	23.2	19.1	7.0	6.1
N 211-T₁ 80/100 (1968)	T.S.	16.7	42.0	29.7	11.6		
	B.P.	14.8	55.1	22.5	7.6	7.2	4.0
N 211-T₂ 80/100 (1968)	T.S.	18.4	33.6	29.9	18.1		
	B.P.	18.2	43.2	23.9	14.7	6.0	3.4
N 211-T₃ 80/100 (1968)	T.S.	20.0	32.6	28.3	19.1		
	B.P.	18.7	45.9	18.7	16.7	9.6	2.4
N 238 (1978)	T.S.	9.7	39.6	33.4	17.3		
	B.P.	15.2	46.0	24.9	13.8	8.5	3.5
N 569 (1975)	T.S.	14.8	32.0	34.5	18.7		
	B.P.	14.9	42.1	28.6	14.4	5.9	4.3
N 579 (1971)	T.S.	15.9	30.7	35.7	17.7		
	B.P.	16.8	43.3	25.9	14.1	9.8	3.6
N 934 (1977)	T.S.	15.2	31.1	34.0	19.7		
	B.P.	12.8	37.5	34.2	15.5	-0.2	4.2
N 972 (1978)	T.S.	13.0	32.5	35.6	18.9		
	B.P.	18.5	34.3	29.7	17.5	5.9	1.4
					Average	6.6	3.6

compounds do not transform directly into asphaltenes, but pass through the "resin" stage before becoming asphaltenes with a molecular configuration that may be different from that of the original asphaltenes.

Thus it can be proposed that the ageing reaction of a bitumen can be as follows in terms of compositional transformation:

$$Cyclics \longrightarrow Resins \longrightarrow Asphaltenes$$

An examination of the differences in the cyclics, resins and asphaltenes contents of the bitumens from the top slices (TS) and bulk portions (BP) of the closed bituminous wearing courses has shown that natural ageing in service is characterized by:

- a rather substantial reduction in cyclics (9% on average over the fifteen sections considered in the study);

- a middle-sized increase in resins (6.6% on average);

- a relatively small increase in asphaltenes (3.6% on average).

THE LCPC EXPERIMENTS

The field ageing studies[2] at the French Laboratorire Central des Ponts et Chaussees on bitumens are recorded in Table II. Results demonstrate that a smaller increase in asphaltenes was observed as compared to resins content. One may, therefore, conclude that natural ageing in the road has occurred due to decrease in cyclics,(which transform mainly into resins), and by a limited increase of 3 to 5% in asphaltenes.

SIMULATION OF BITUMEN AGEING IN THE LABORATORY

The study of bitumen ageing in service (Table I) has made it possible to extract and qualitatively recover a number of binders from the bulk portions (BP) of wearing courses having a low percentage of voids(e.g. less than 5%). It can be presumed that, being shielded from air and moisture, this bitumen still has the same chemical composition at the time of construction of the pavement, (i.e., as after the laying and compaction of the bitumen mix. The objective of the simulation experiment will, therefore, consist of trying artificial ageing conditions so as to have those bitumens develop into a composition similar to that of the aged bitumens in the corresponding top slices (TS).

The oxidation of bitumen in different ageing tests has been discussed by previous investigators[3]. These authors concluded that among the eight tests studied, the Thin Film

Oven Test and the Rolling Thin Film Test with air flow
yielded the best correlation with ageing in practice during
the manufacture of the mix up to and including the
compaction of the layer.

In the Germany, it was proposed that an ageing
resistance tests at 180^0C to be conducted for one hour on
bitumen-coated aggregate in the presence of 500 mL/min flow
of air. Peterson et al.[5] had performed the ageing tests at
130^0C on bitumen-coated aggregate placed in a
chromatographic column through which air was flowed.

AGEING CONDITIONS

The accelerated bitumen ageing is dependent upon a
number of factors which stimulate oxidation and the kinetics
of the reaction. These factors include mainly:

- rise in temperature;

- composition of the ambient atmosphere (inert, air,
  pure oxygen);

- renewal of the bitumen surface in contact with this
  atmosphere, which cannot be achieved if the binder is
  not sufficiently fluid.

The ageing conditions used were as follows:

- the tests were carried out in a Bucho-type evaporator
  fitted with a one-liter flask which turns in an oil
  bath;

- the mass of bitumen per test was 240 to 250 g;
- temperature was 130^0C;

- the rotation of the flask was ten revolutions/minute;

- the atmosphere above the bitumen was pure oxygen
  flowing at a rate of 4-5 liters/hour as measured with
  a rotameter);

- ageing times were eight and twenty-four hours.

About 240-250 g of bitumen extracted and recovered from the
bulk portions of the wearing courses in twelve of the
fifteen sections mentioned in Table I was subjected to
accelerated ageing under the experimental conditions. After
8 and 24 hours, a bitumen sample was collected for
composition analysis by chromatography.

DISCUSSION

For each of the three families of compounds involved in
the ageing process (cyclics, resins and asphaltenes shift),
Table III shows the differences in contents found after
ageing in the road and after 8 and 24 hours of accelerated

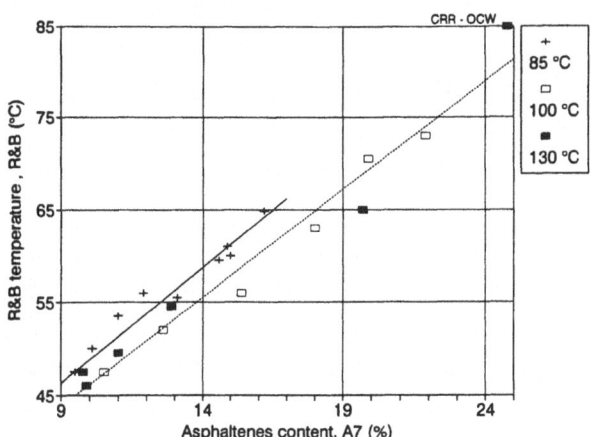

Figure 1. Ring and Ball Temperature Versus Asphaltenes Content.

Table II. Variations in the Resins and Asphaltenes Contents of Three in Service Aged Bitumens (*).

	Asphaltenes A7 (%)			Resins (%)		
	Bit. A	Bit.B	Bit.C	Bit.A	Bit.B	Bit.C
Original bitumen	17.7	10.5	13.6	19.3	23.4	17.2
1986	19.6	12.4	15.3	19.3	22.8	25.0
1987	20.6	15.7	18.8	24.6	25.1	20.2
1988	22.9	15.1	18.6	28.4	30.0	34.2
1989	22.9	14.6	20.0	35.2	30.1	31.6
1990	22.8	13.8	18.8	29.9	32.0	33.1
Δ	5.1	3.3	5.2	10.6	8.6	15.9

(*) These values have been obtained from reference (2).

Table III. Development of Generic Compositions after Ageing in the Road (Field) and in the Laboratory (Lab. Ageing at 130°C and O_2 Flow).

	T1 80/100	T2 80/100	T3 80/100	T1 50/70	T2 50/70	T3 50/70	N569	N53	N49	N27	N238	RO	Aver.
Δcyclics (%) field ageing	-13.1	-9.6	-13.3	-16.6	-13.9	-10.7	-10.1	-6.6	-7.9	-2.3	-6.4	-3.2	-9.5
Δcyclics (%) after 8 h lab. ageing	-10.3	-1.4	-9.1	-6.2	-4.4	-3.1	-4.6	-4.0	-8.3	+4.0	-6.0	-3.1	-4.7
Δcyclics (%) after 24 h lab. ageing	-13.3	-6.4	-12.9	-11.3	-10.6	-5.8	-4.8	-5.1	-12.2	+0.5	-12.8	-7.9	-8.5
Δresins (%) field ageing	+7.2	+6.0	+9.6	+10.4	+10.5	+7.0	+5.9	+9.6	+4.7	+4.3	+8.5	-0.6	+6.9
Δresins (%) after 8 h lab. ageing	+9.4	-0.8	+4.0	+3.2	+1.6	-4.5	-4.5	+4.3	-1.8	-3.8	+2.5	+1.5	+0.9
Δresins (%) after 24 h lab. ageing	+7.6	+0.8	+2.3	+4.7	+1.0	-1.9	-1.4	+5.9	-0.9	-0.8	+1.9	-0.1	+1.7
ΔA7 (%) field ageing	+4.0	+3.4	+2.4	+6.7	+4.5	+6.1	+4.3	+1.9	+4.9	+2.4	+3.5	+1.5	+3.8
ΔA7 (%) after 8 h lab. ageing	+3.4	+1.1	+4.8	+3.3	+3.4	+4.1	+5.5	+3.3	+3.7	+6.7	+5.2	+4.7	+4.1
ΔA7 (%) after 24 h lab. ageing	+6.7	+5.4	+9.1	+7.1	+7.5	+4.9	+5.1	+4.3	+16.9	+7.9	+12.5	+11.6	+8.3

ageing. More resins were formed in the road than that during 8 and 24 hours of accelerated ageing. The average values recorded for the twelve investigated sections were 6.9% of the resins formed in service and only 0.9% and 1.7% produced after 8 and 24 hours in the laboratory. Relatively few asphaltenes were formed in the road. After 8 and 24 hours of treatment at 130^0C, the asphaltenes content had increased more than that in natural ageing in road. Therefore, it appears that ageing in the road is characterized mainly by the formation of resins, owing to the much more moderate conditions of temperature and atmosphere than in accelerated ageing.

It appears that 130^0C is too high to simulate natural ageing and favors the formation of asphaltenes. On average over the twelve investigated sections 3.8% of asphaltenes were produced in service and 4.1 and 8.3% of asphaltenes were obtained at 130^0C under oxygen environment. It is recommended that the natural ageing should be simulated under less severe conditions of temperature.

REQUIRED CONDITIONS FOR LABORATORY AGEING

A previous study[6] related to developments in the generic compositions of bitumens has demonstrated that the following conditions should be met in simulating field ageing:

- work at a selected temperature between 80 to 90^0C;

- use an atmosphere of pure oxygen (flow of 4-5 liter/hour), and

- conduct the test for five to six days (144 hours).

The validity of these conditions have been confirmed by an ageing test performed on a 80/100 pen. bitumen (569) at three different temperatures 85, 100 and 130^0C with ageing times adapted to the selected temperature[7]. The ring and ball temperature, asphaltenes content and migration of the material were monitored for the three ageing temperatures. The data obtained are recorded in Table IV.

It appears from the results that the formation of asphaltenes is significantly lower at 85^0C during 168 hours ageing. These observations were similar to asphaltene formation (5 to 6%) in the natural ageing process.

Moreover, when confronting asphaltene content with ring and ball temperature or the spot test or the reciprocal of penetration(1/pen.), respectively, a different relation is found according to whether the test was performed at 85^0C on the one hand or at 100 or 130^0C on the other, which is usually not normal (Figures 1 and 2). The choice of temperature in laboratory ageing is, therefore, an important one. The conditions recommended above (85^0C) are valid and

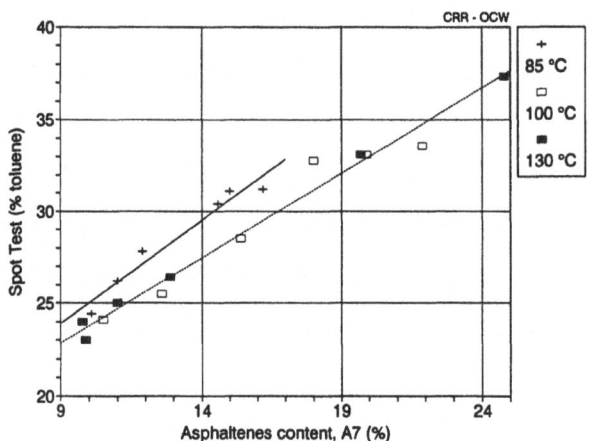

Figure 2. Spot Test Value Versus Asphaltenes Content.

Table IV. Influence of Ageing Temperature on Penetration, R&B and Asphaltenes Content.

t (hours)	Ageing T° 85°C			Ageing T° 100°C			Ageing T° 130°C		
	pen mm/10	R&B °C	A7 %	pen mm/10	R&B °C	A7 %	pen mm/10	R&B °C	A7 %
0	84	44.5	9.6	84	44.5	9.6	84	44.5	9.6
1	-	-	-	-	-	-	81	46.0	9.9
2	-	-	-	-	-	-	66	47.5	9.8
4	-	-	-	62	47.5	10.5	58	49.5	11.0
8	54	50.0	10.1	51	52.0	12.6	42	54.5	12.9
24	45	53.5	11.0	36	56.0	15.4	21	65.0	19.7
48	-	-	-	-	-	-	11	85.0	24.8
72	37	56.0	11.9	22	63.0	18.0	-	-	-
168	28	59.5	14.6	17	70.5	19.9	-	-	-
240	-	-	-	15	73.0	21.9	-	-	-
336	26	60.0	15.0	-	-	-	-	-	-
504	22	64.5	16.2	-	-	-	-	-	-

that temperatures of 100^{0}C and above result in too rapid an ageing which does not reflect the process as it actually occurs in a road.

The conditions recommended for laboratory ageing (85^{0}C, six days, oxygen flow) were applied to a number of bitumens including four bitumens from the binder courses ("bulk portion") of four road sections mentioned in Table I as well as five other 80/100 pen. bitumens.

The results are presented in Table V. These data suggest that under these conditions laboratory ageing is similar to in-service ageing and the formation of asphaltenes remains limited and comparable to that which occurs in natural ageing.

It should also be noted that the values of penetration and ring and ball temperature (not mentioned in Table V) found on the laboratory-aged bitumens agree with those measured on the naturally aged bitumens.

Figure 3 represents the development of the asphaltene content of bitumen 569 during accelerated ageing at three different temperatures. It clearly illustrates that at temperatures of 100^{0}C and above too much asphaltenes are formed for the test to reflect natural ageing in a road.

FORMATION OF ASPHALTENES

Asphaltenes are generally defined as the bitumen fraction which is insoluble in hot n-heptane. These are analyzed by the IP 143/57 procedure, which is rather time-consuming (36 hours), and the accuracy of measurements is ±10% in relative terms. In order to overcome these difficulties, a method called the "SPOT TEST" has been developed with an accuracy of ± 1% and a shorter time of execution (20 minutes).

The spot test procedure is described as follows: 0.5 g bitumen was dissolved in 10 mL toluene in a 100 mL conical flask using magnetic stirrer. The dissolved bitumen was titrated with n-heptane. Upon each addition of 1 mL of n-heptane a drop of the mixture was collected, and placed it on a filter paper. The addition of n-heptane was continued until two concentric stains appear on the filter. The volume (V) of the added n-heptane was recorded. The results were expressed in toluene % as follows: 10/(10 + V) x 100. The test was performed in duplicate using samples, but the trial for two stains appearance was performed when the added volume was (V - 1) mL and thereafter upon each addition of 0.1 mL of n-heptane. The reported spot test values are the mean of three readings.

Table VI lists the data on bitumen 569 for three ageing temperatures (85, 100 and 130^{0}C). The results of the spot test can be described as follows:

Table V. Comparison Between in the Field Aged Bitumen and in the Laboratory Aged (144 hours at 85°C under O_2 Flow) Bitumen.

Bitumen	% A7 Original bitumen	% A7 Field	ΔA7 % Field	% A7 Laboratory	ΔA7 % Laboratory
T1 50/70	7.1	13.8	6.7	13.4	6.3
T3 50/70	19.1	25.2	6.1	21.1	2.0
N 49	11.7	16.6	4.9	15.9	4.2
N 569	14.4	18.7	4.3	19.8	5.4
Bit. 577	9.2	-	-	12.8	3.6
Bit. 569	9.6	-	-	14.6	5.0
Bit. 467	10.8	-	-	15.5	4.7
Bit. 472	10.7	-	-	14.7	4.7
Bit. 481	9.4	-	-	13.5	4.1

Table VI. Influence of Ageing Temperature on Spot Test Value and Asphaltenes Content.

t (hours)	Ageing T° 85°C V * mL	Ageing T° 85°C ** %	Ageing T° 85°C A7 %	Ageing T° 100°C V * mL	Ageing T° 100°C ** %	Ageing T° 100°C A7 %	Ageing T° 130°C V * mL	Ageing T° 130°C ** %	Ageing T° 130°C A7 %
0	49.2	16.9	9.6	49.2	16.9	9.6	49.2	16.9	9.6
1	-	-	-	-	-	-	33.5	23.0	9.9
2	-	-	-	-	-	-	31.6	24.0	9.8
4	-	-	-	31.5	24.1	10.5	29.9	25.0	11.0
8	31.0	24.4	10.1	29.1	25.5	12.6	27.8	26.4	12.9
24	28.2	26.2	11.0	25.0	28.5	15.4	20.2	33.1	19.7
48	-	-	-	-	-	-	16.8	37.3	24.8
72	25.9	27.8	11.9	20.5	32.8	18.0	-	-	-
168	22.8	30.5	14.6	20.2	33.1	19.9	-	-	-
240	-	-	-	19.7	33.6	21.9	-	-	-
336	22.2	31.0	15.0	-	-	-	-	-	-
504	22.0	31.2	16.2	-	-	-	-	-	-

* V = mL of n-heptane   ** : Spot test value (% of toluene)

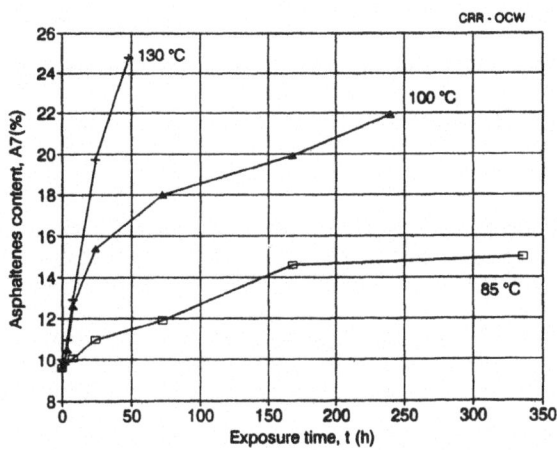

Figure 3. Influence of Ageing Temperature on the Development of Asphaltenes Content of Bitumen.

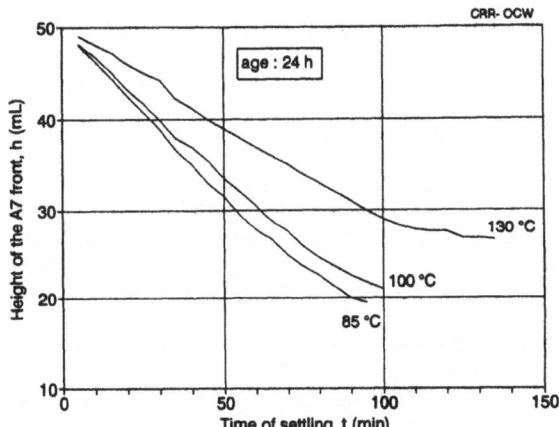

Figure 4. Influence of Ageing Temperature on Settling Rate of Asphaltenes.

- the spot test is easy to perform and does not require sophisticated experimental design and equipment.

- the spot test may be an alternative to the classical determination of asphaltenes (precipitation and separation in hot n-heptane)

- it allows rapid and sensitive monitoring of the development of asphaltenes in an ageing process

- when comparing with the asphaltenes found by the usual method, this procedure indicates (Figure 2) that the asphaltenes formed at 85°C in accelerated ageing are different from those formed at 100 and 130°C.

In 1979 Plancher and co-workers[8] used an asphaltenes settling test in hexane to study the influence of laboratory ageing time (ASTM D2782: Rolling Thin Film Oven Test - RTFOT) on settling rate. This test was also performed on eight bitumens of different penetration grades and no correlation was found between settling rate on the one hand and asphaltenes content and penetration on the other. Plancher and co-workers generally found a reduction in settling rate with an increase in RTFOT ageing time.

The settling test performed in n-heptane using bitumen 569 for three ageing temperatures (85, 100 and 130°C) and the different laboratory ageing times. The data for the bitumen investigation can be described as follows:

- for a given accelerated ageing temperature a reduction in settling rate with an increase in ageing time;

- for a given accelerated ageing time a decrease in settling rate with an increase in ageing temperature (Figure 4). According to Plancher this is due to the fact that at low temperature (85°C) the asphaltenes are spherical whereas at high temperatures the shape of asphaltene is like flakes that are subject to a "parachute" effect when settling down.

Comparing with the first set of data, the settling test would be worth considering further in a more thorough study including among other parameters in an application to field-aged bitumens in order to explain the followings: Is the decrease in settling rate with increasing duration of the ageing test due to the peptization of the micelles as a result of increased polarity? or should it rather be ascribed to be shape or the fragmentation of the asphaltenes?

A KINETIC APPROACH TO AGEING

A study[9] conducted at the Belgian Road Research Centre in 1991 has shown that among the various classical kinetic models and those for solid-state reactions, generally that of one-dimentional diffusion (n=2 in the equation below) best applies to tests performed with selected indicators (S): asphaltenes content, ring and ball softening temperature, and penetration (1/pen.).

The corresponding equation is as follows:

$$a^n = [(S_t - S_0)/(S_f - S_0)]^n = (Ds_t/Ds_\infty)^n = kt$$

where a =   extent of reaction
   S_0 =   value of the indicator being considered at time t = 0,
   S_t =   value of indicator being considered at time t,
   S_f =   value of indicator being considered at time t = t_{final}.

By rearranging this equation:

$$S_t = S_o + K^{1/n} \cdot t^{1/n}$$

Hence:

$$K = DS^n \cdot k$$

where K = overall reaction rate.

Using these equations, the exponent is determined which most accurately fits the results obtained in accelerated ageing tests conducted at different temperatures on various bitumens. The value of n seems to be 3/2 for ageing temperatures of 100°C, whichever indicator is considered.

Various cases were examined and, generally speaking, exponent n was often close to 3/2 for temperatures above 100°C (n value giving the best or a very high degree of correlation). Although the physical meaning of this exponent 3/2 is not clear as far as the proposed kinetic approach is concerned, this exponent makes it possible to mathematically process results obtained at temperatures above 100°C and to represent the development of the binder's characteristics with time.

The kinetic approach being applied to the results on bitumen 569 (Table IV), the exponent n giving the best degree of correlation proves to be either 2 or 3/2, depending on the indicator under consideration, when the ageing temperature is 100°C. This temperature of 100° seems to fall into the range of transition between two different ageing processes. This observation and the kinetic approach of bitumens ageing have already been reported previously[6,9,10].

CONCLUSIONS

It is concluded from the observations of the present study that the natural ageing causes only a small increase in asphaltenes content. In general, the bitumen ageing in a road is characterized by the following sequence:

cycles $\longrightarrow$ resins $\longrightarrow$ asphaltenes

Accelerated "in service" ageing tests in the laboratory should be conducted at temperature below $100^{0}C$ (e.g., between 70 and $95^{0}C$). An appropriate temperature for comparison purposes seems to be $85^{0}C$. Many accelerated ageing tests carried out have shown that 144 hours of laboratory ageing at $85^{0}C$ under O_2 flow are equivalent to about twenty years of field ageing. When relating asphaltenes content and other technological characteristics (ring and ball softening temperature, penetration, 1/pen., spot test), it appears that the mechanism of asphaltenes formation and ageing in a bitumen is different according to whether this bitumen has aged at a temperature below or above $100^{0}C$. A kinetic approach to accelerated ageing has confirmed that $100^{0}C$ is a critical temperature at which the ageing mechanism or process seems to change.

REFERENCES

1. Choquet, F., The Development of Bituminous Binders in Pavements in Service, 4th Eurobitume Symposium, Vol. I, Summaries and Papers, pp. 30-34, Madrid (1989).

2. Farcas, F. and Such, C., Influence de la Structure Chimique des Bitumes sur Leur Comportment Rheologique: Description des Methodes Analytiques Utilisees; Journee d'etude AFREM (LCPC) - Les Bitumes, Leur Fabrication, Leurs Proprietes et Leurs Applications, pp. 29-55, Saint-Remy-les-Chevreuse (France), 28-29, Novembre (1991).

3. Van Gooswilligen, G., de Bats, F. Th., and Berger, H., Oxidation of Bitumens in Various Tests, 3rd Eurobitume Symposium, Vol. I, pp. 95-101, The Hague (1985).

4. Potschka, V., Alterungsbestandigkeit und Homogenitat von Bitumen, Strasse und Verkehr 2000, Vol. 2B-21, pp. 285-288, Berlin (1988).

5. Davis, T. C. and Peterson, J. C., An Adaption of Inverse Gas-Liquid Chromatography to Asphalt Oxidation Studies, Analytical Chemistry, Vol. 38, No. 13, pp. 1988-1190 (1966).

6. Choquet, F. S., The Search for an Ageing Test Based on Changes in the Generic Composition of Bitumens, International Symposium on "Chemistry of Bitumens", Proceedings, Vol. II, pp. 787-812, Rome (1991).

7. Meuleman, F., Comparison entre Methodes de Vieillissement Accelere du Bitume, Travail de fin d'etudes, Universite Libre de Bruxelles, Annee Academique,p. 108, Centre de Recherches Routieres, Bruxelles, Belgium, (1991/1992).

8. Plancher, M., Hoiberg, A. J., Suhaka, S. C. and Peterson, J. C., A Settling Test to Evaluate the Relative Degree of Dispersion of Asphaltenes, Asphalt Paving Technology, Vol. 48, pp. 351-374 (1979).

9. Verhasselt, A. F. and Choquet, F. S., A New Approach to Studying the Kinetics of Bitumen Ageing, International Symposium on "Chemistry of Bitumen", Proceedings, Vol. II, pp. 686-705, Rome (1991).

10. Verhasselt, A. F., A Kinetic Approach to the Ageing of Bitumens, In "ASPHALTENES AND ASPHALT", Vol. II, Development in Petroleum Science Series, (Editors - T. F. Yen and G. V. Chilingarian), Elsevier Science Publishers, The Netherlands, To be Published.

FIELD-SCALE NUMERICAL SIMULATION OF

BITUMEN MOBILIZATION WITH SOLVENT SLUGS

J.M. Carter and M.R. Islam

Department of Geological Engineering
South Dakota School of Mines and Technology
Rapid City, SD 57701 (USA)

ABSTRACT

A comprehensive numerical simulator is developed from an existing bitumen leaching model to incorporate the effects of asphaltene precipitation during bitumen mobilization with a solvent. Such mobilization is required prior to any thermal enhanced oil recovery (EOR) application. Numerical simulation results indicate that the presence of a bottom-water zone, often considered to be detrimental to oil production, enhances bitumen mobilization to a great extent. The bottom-water zone acts as a transporting medium for initial movement of the solvent and invokes solvent leaching of bitumen. Asphaltene precipitation reduces permeability of the bottom-water zone, thus, controlling excessive fluid influx into the bottom-water zone. This role of asphaltene is particularly important for relatively higher bottom-water permeabilities. A series of numerical simulation runs was conducted to study the effect of a wide range of parameters, such as, the bitumen-to-water zone permeability ratio, type of solvent, the bitumen-to-water zone thickness ratio, flow rate, and others.

INTRODUCTION

Enhanced oil recovery (EOR) methods for bitumen recovery are becoming increasingly important to the United States and Canada as the amount of light oil is decreasing and the political climate in the Persian Gulf remains unstable. Currently, the world's EOR and heavy oil production is approximately 1.9 million bbl/day, which is 3.2% of the average 1991 oil production rate of 59.96

Asphaltene Particles in Fossil Fuel Exploration, Recovery, Refining, and Production Processes, Edited by M.K. Sharma and T.F. Yen, Plenum Press, New York, 1994

31

million bbl/day. In fact 10% of the United States oil production is from enhanced oil recovery projects[1], which indicates the effectiveness of EOR methods.

Only seven percent of the oil sand reserves can be exploited by surface mining[2], the rest has to be produced by in situ methods. The biggest problem in producing such a high-viscosity fluid is the low mobility. The most effective way to reduce mobility is the use of heat. Steam can be used to provide heat to the oil sands. Even though, the reduction of viscosity through heat injection is the quickest way to reduce bitumen viscosity, one of the biggest problem is to achieve mobility during initial steam injection in the cold bitumen formation. Low injectivity of steam, especially during initial stages of injection, makes it extremely difficult to carry on the injection process.

Solvent injection has been proposed to decrease oil viscosity[3-6]. The advantage of solvent is in the fact that a high temperature is not required to start up the process. Solvents can effectively reduce oil viscosity with a diffusion process under cold reservoir conditions. However, it is very expensive to use solvent throughout the recovery process. Besides, solvent dilution is an extremely slow process due to its dependence on diffusion alone. The process can be enhanced substantially by injecting solvent followed by steam. This process of steam injection with solvent slug has gained popularity in recent years[2,7,8]. Even though the process has not been applied in a field, the process seems to have great potential in future.

Recently, Islam[9], and Islam and Farouq Ali[10] have discussed the potential of solvent slug followed by steam in recovering heavy oil from bottom-water reservoirs. In the past, the presence of a bottom-water zone had been considered to be detrimental to oil production. Proctor et al.[8] are the first group to report that the presence of a bottom-water zone can in fact increase oil production if the bottom-water zone can be used effectively as the transporting medium. These high-saturation zones may be beneficial in providing initial injectivity and also in providing a flow path for the mobilized bitumen. Murji[2] proposed the use of solvent followed by steam in heavy oil reservoirs containing a bottom-water zone. Scaled model results showed significant recovery with solvent alone.

This paper presents numerical simulation of bitumen recovery with solvents. The proposed simulator is the first step in developing a comprehensive mathematical model of solvent followed by steam process.

NUMERICAL SIMULATION

The numerical simulation, presented here, is a modified version of the isothermal, two-dimensional, two-phase, two-component model to a solvent-bitumen system[2,11]. The simulator is modified to incorporate the effect of asphaltene precipitation and asphaltene blockage. Following is a description of the numerical simulation scheme.

The instantaneous liquid saturation, S, is given by

$$S = \frac{\phi}{\phi_m} \tag{1}$$

where, ϕ is the volume of the flowing liquid in which convection can occur and ϕ_m is the volume fraction void of sand in which both bitumen and solvent phases coexist. If the spherical solute particles have a uniform local size distribution with a particle number density of n_p and an average particle radius of r_p, macroscopic liquid saturation may be expressed by

$$S = 1 - \frac{4\pi}{3} \frac{n_p r_p^3}{\phi_m} \tag{2}$$

The particles shrink as bitumen is desorbed, and the velocity of dissolution, $\dot{r}_p$, is given by the radial advance of contact, as follows

$$\dot{r}_p = \frac{dr_p}{dt} \tag{3}$$

The radial distribution integro-differential equation for the sphere radius, r_p, as a function of the macroscopic liquid phase solvent concentration, C, is written as follows

$$C^* \frac{\partial}{\partial t} \left[r_o F(t) + \int_0^t \dot{r}_p(\tau) F(t-\tau) \, d\tau \right] = r_p^2 \left[C \dot{r}_p + \alpha \left(C - C^* \right) \right] \tag{4}$$

where

$$F(t) = r_p^2 \left[\frac{1}{3} - \frac{2}{\pi^2} \sum_{k=1}^{\infty} e^{-k^2 \pi^2 D_s t / r_p^2} \right] \tag{5}$$

Longitudinal and transverse coefficients of dispersion are given by

$$D_t = \frac{D_o}{F_R} + 0.0157 \sigma d_p |\mathcal{V}| \tag{6}$$

$$D_l = \frac{D_o}{F_R} + 0.5 \sigma d_p |\mathcal{V}| \tag{7}$$

Equating the rate of solvent accumulation to the combined convective and diffusive fluxes gives the convective-diffusion-adsorption equation as follows:

$$\frac{\partial}{\partial X_i}\left[D_{ij}\frac{\partial C}{\partial X_j}-V_iC\right]=\left[\phi_m-\frac{4\pi}{3}n_pr_p^3\right]\frac{\partial C}{\partial t}+4\pi n_pr_p^2\left[\alpha\left(C-C^*\right)\right] \qquad (8)$$

The flow equation is given by Darcy's law as follows

$$\overline{V}=\frac{\overline{u}}{\rho_s C} \qquad (9)$$

$$\overline{V_i}=-\frac{K_i}{\mu}\frac{\partial}{\partial X_i}\left(p+\rho gh\right) \qquad (10)$$

The Darcy continuity equation is given by

$$\frac{\partial}{\partial X_i}\left[\frac{\rho K_i}{\mu}\frac{\partial}{\partial X_i}\left(p+\rho gh\right)\right]=\phi_m\left[1-\frac{4\pi}{3}n_pr_p^3\right]\frac{\partial p}{\partial t} \qquad (11)$$

The Kozeny-Carman equation is used to analytically relate the component permeabilities to porosity. The bitumen-solvent viscosity is related to the liquidity determined by the solvent component viscosities and densities using Cragoe's method[6].

Asphaltene precipitation is considered to be a function of solvent concentration as well as bitumen saturation. The current modeling approach does not allow the movement of asphaltene. As the concentration goes up in a certain block, asphaltene precipitation is activated if the saturation of the block exceeds predetermined value. The table provided by Kamath et al.[12] was used to generate the precipitation versus concentration relationship. As asphaltene precipitation occurs, the porosity, as well as permeability, of the porous medium decreases. The decrease in permeability is considered as outlined by Islam and Farouq Ali[10]. They expressed local permeability as:

$$\frac{k_x}{k_i}=1-\frac{\beta\rho}{\phi_i} \qquad (12)$$

Also, for steady state:

where ß is the flow restriction parameter which is considered to be a function of the pore throat diameter. In the present study, we consider ß to be a function of absolute permeability of the porous medium as well as the

$$\frac{k_\infty}{k_i} = 1 - \frac{\beta}{\alpha} \qquad (13)$$

average asphaltene particle size. Table 1 gives ß values for different permeabilities.

NUMERICAL SOLUTION

The numerical scheme has been detailed by Oguztoreli and Farouq Ali[10]. The approach is described here briefly. The radial dissolution equation (Eq. 4) is discretized using central differences for the time domain and the midpoint rule for integration. The Darcy-Continuity equation (Eq. 11) is discretized using implicit pressure formulation. On the other hand, Eq. 8 is discretized by using central difference in space and Crank-Nicholson scheme in time. The discretized sets of algebraic equations are solved by direct method. The mid-point velocities and diffusion coefficients are evaluated at the mid points from the current pressure iterate to generate values for the concentration at the next consecutive node. In order to model the presence precipitated asphaltene, the procedure outlined by Islam and Farouq Ali[10] was used. This involves recalculation of the permeability terms according to the amount of asphaltene present in the porous medium.

RESULTS AND DISCUSSION

The model dimensions used for the simulation were 300 ft length, 200 ft width, and a 50 ft total bitumen and bottom-water zone thickness. Figure 1 shows geometry and dimensions of the field modeled in this paper. Table 1 shows all the relevant data for different numerical simulation runs.

The first parameter that was studied was the ratio of the bitumen to bottom-water zone permeability. Water zone-to-bitumen zone permeability ratios of 1, 2, and 4 were modeled. Results are shown in Figure 2. For all these numerical runs, a bitumen-to-water zone thickness ratio of 2.25 was used. Figure 2 shows that cumulative bitumen recovery increases as the permeability ratio increases. When this ratio is equal to unity, only 8% of the initial bitumen in place is recovered even when the solvent is injected for 1.7 hydrocarbon pore volumes (HCPV). Solvent breakthrough takes place after the injection of about 0.3 PV. Following solvent breakthrough, practically no additional bitumen is produced. As the permeability ratio is increased (i.e., as the bottom-water zone becomes tighter), the bitumen recovery increases drastically. The permeability ratio of 4 yielded a total recovery of 40% of the bitumen in place. For a permeability ratio higher than 4, injection pressure increased substantially without any improvement on bitumen recovery. Consequently, a permeability ratio of 4 was considered to be optimum.

TABLE 1. CHARACTERISTICS OF NUMERICAL RUNS

Run no.	Perm. ratio (k/k)	Thickness ratio (h/h)	Flow rate(m)	Solvent viscosity	Diffusion coef. (cm/day)
Run 1	1	2.25	20	4.56	.5 10
Run 2	2	2.25	20	4.56	.5 10
Run 3	4	2.25	20	4.56	.5 10
Run 4	1	2.25	10	4.56	.5 10
Run 5	2	2.25	40	4.56	.5 10
Run 6	2	2.25	60	4.56	.5 10
Run 7	2	2.25	20	4.56	.5 10
Run 8	2	2.25	20	2.28	.5 10
Run 9	2	3.33	20	1.14	.5 10
Run 10	2	5.5	20	4.56	.5 10
Run 11	2	12	20	4.56	.5 10
Run 12	2	2.25	20	4.56	1. 10
Run 13	2	2.25	20	4.56	.25 10
Run 14	2	2.25	20	4.56	.1 10
Run 15	2	12	20	4.56	.5 10
Run 16	2	2.25	20	4.56	.5 10
Run 17	4	2.25	20	4.56	.5 10
Run 18	1	2.25	20	4.56	.5 10

asphaltene precipitation considered

Flow restriction parameter, β	Absolute permeability (μm) β
14	13
7	6.5
3.5	5

Reservoir Characteristics

Porosity	35%
Permeability	14 μm
Bitumen viscosity	190,000 ma.s
Water viscosity	1 ma.s
Initial water saturation	25% bitumen zone
	100% water zone

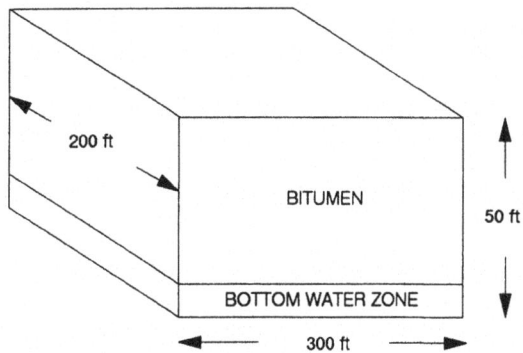

Figure 1. Geometry and dimensions of the field prototype

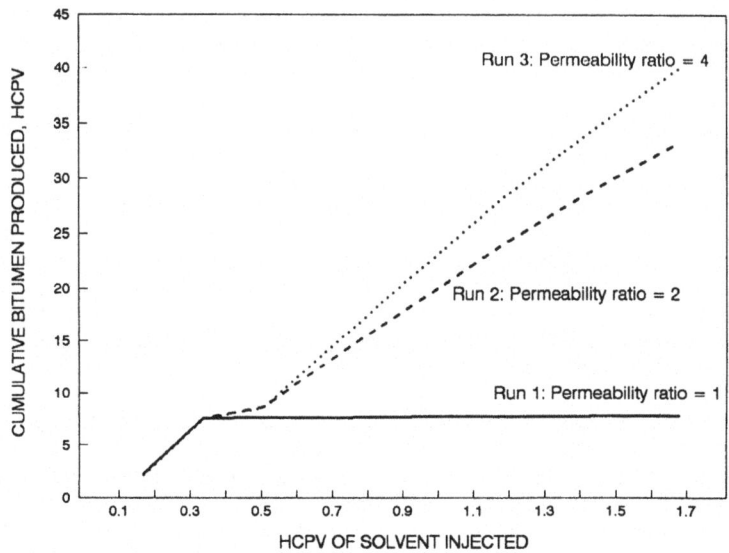

Figure 2. Effect of bitumen-to-water zone permeability ratio on bitumen recovery

The second parameter that was studied was the solvent injection rate. The effects of flow rate was studied through a series of numerical runs at flow rates of 10, 20, 40, and 60 m^3/day. For all these runs, a bitumen-to-water zone thickness ratio of 2.25 was used. All other characteristics are listed in Table 1. Figure 3 shows results of different runs conducted at different solvent injection rates. This figure shows that a slower injection rate leads to an increase in recovery. A slow injection rate improves recovery by diffusion and enhanced leaching. However, rate of return is rather small with a slow recovery process and the economics of such a process may be prohibitive. This aspect is shown in Figure 4 which shows cumulative recovery as a function of time for different flow rate cases. As can be seen from this figure, it would take approximately 11 years to produce 42% of the bitumen at the lower flow rate. However, it would take only one and a half years to produce 24% of bitumen at the higher flow rate.

The third parameter studied in this paper is solvent viscosity. Viscosities of 4.56 cp, 2.28 cp, and 1.14 cp were modeled. For all these runs a bitumen-to-water zone thickness ratio of 2.25 and a permeability ratio of 2 were used. Results of runs with solvent viscosities of 4.56 cp and 2.28 cp are shown in Figure 5. Results with a lower solvent viscosity was not any different from the 2.28 ma.s case. Figure 5 shows that the bitumen recovery remains the same for both solvent viscosities. However, recovery is improved later for the case of lower solvent viscosity. This trend of increasing recovery with decreasing solvent viscosity does not continue for a solvent viscosity below 2 ma.s.

The fourth parameter modeled in this study is the ratio of bitumen layer thickness to the bottom-water zone thickness. Thickness ratios of 5.5, 12, 3.3 and 2.25 were studied. All other parameters of these runs remain the same. They are listed in Table 1. Figure 6 shows recovery curves for all these runs. As can be seen from this figure, there is no consistent increase in recovery with increasing bitumen-to-water zone thickness ratio. However, the highest recovery is obtained for a thickness ratio of 12.

Molecular diffusion coefficient was the fifth parameter studied in this paper. Diffusion coefficients of 0.1×10^{-5}, 0.25×10^{-5}, 0.5×10^{-5}, and 1.0×10^{-5} cm^2/day were modeled. All other parameters of these runs were kept constant and are listed in Table 1. Figure 7 compares recovery curves for different diffusion coefficient values, Note that as the diffusion coefficient is increased from 0.1×10^{-5} to 0.25×10^{-5} cm^2/day, there is drastic increase in recovery during later stages of recovery. This incremental recovery does not continue beyond 0.5×10^{-5} cm^2/day. Also, recovery is not sensitive to diffusion coefficient values during the period prior to solvent breakthrough. This behavior is expected because the flow during this period is dependent entirely on the injection rate. During later stages, however, diffusion coefficient plays an important role in bitumen leaching which in turn contributes to bitumen recovery.

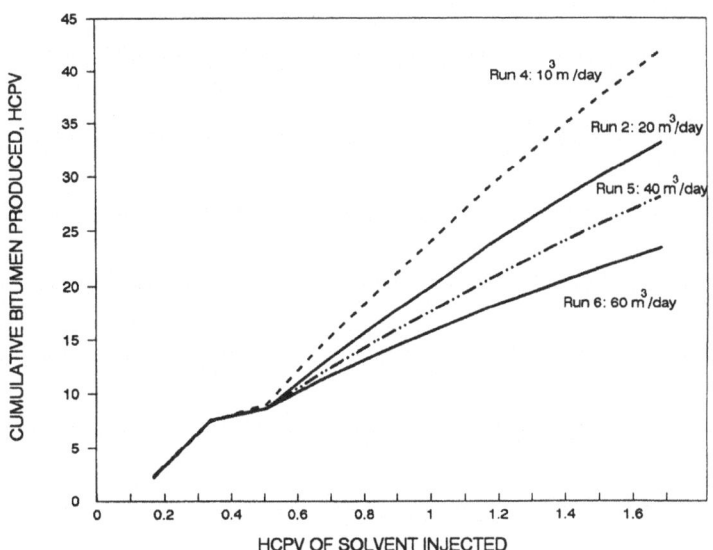

Figure 3. Effect of solvent injection rates on bitumen recovery

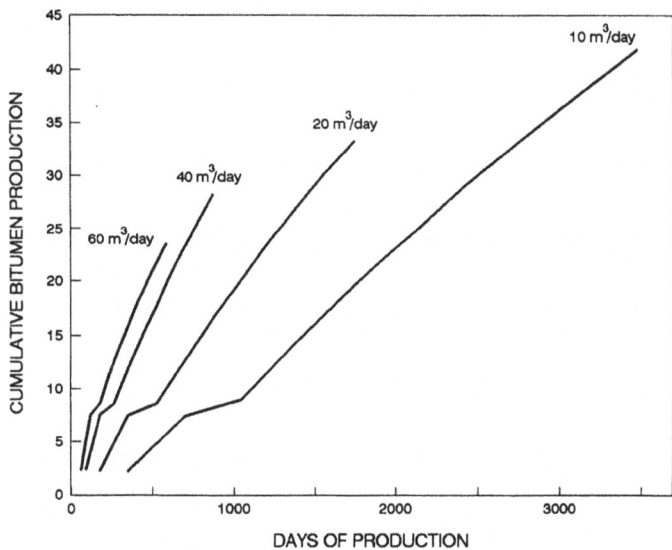

Figure 4. Effect of solvent injection rates on bitumen recovery as a function of time

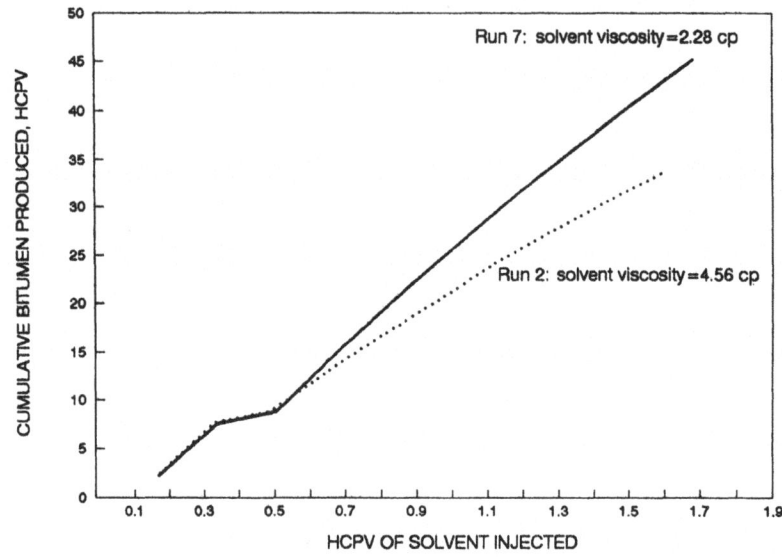

Figure 5. Effect of solvent viscosity on bitumen recovery

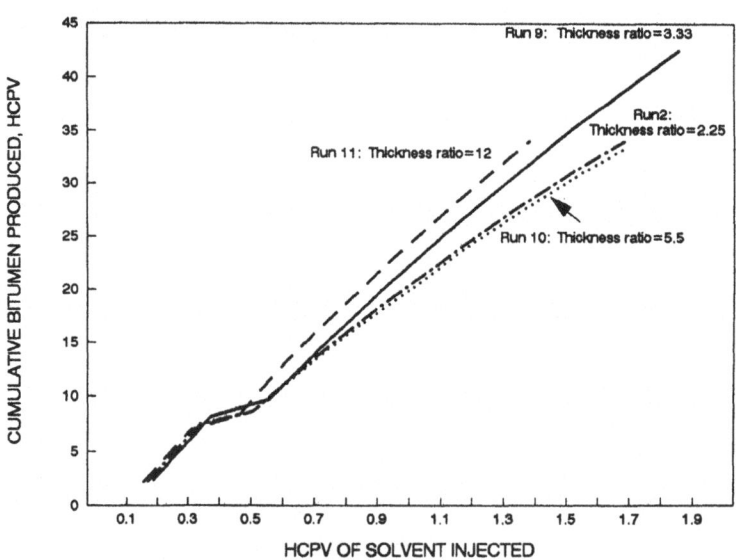

Figure 6. Effect of bitumen-to-water zone thickness ratio on bitumen recovery

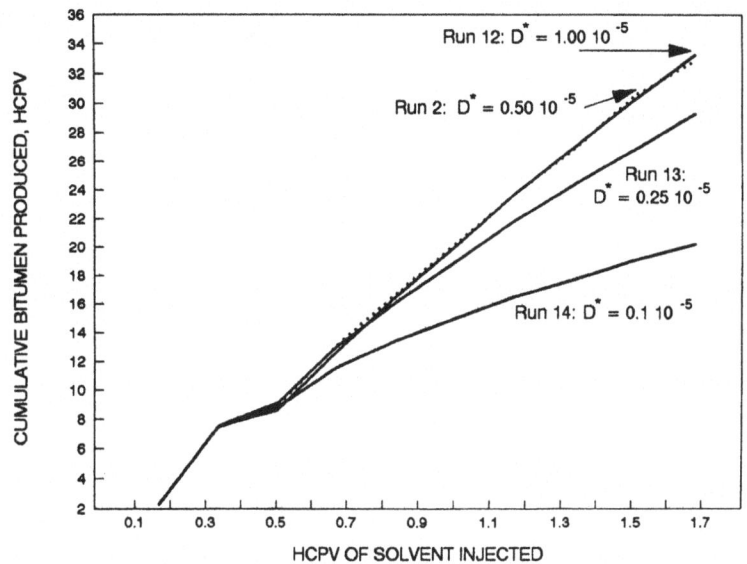

Figure 7. Effect of molecular diffusion coefficients on bitumen recovery

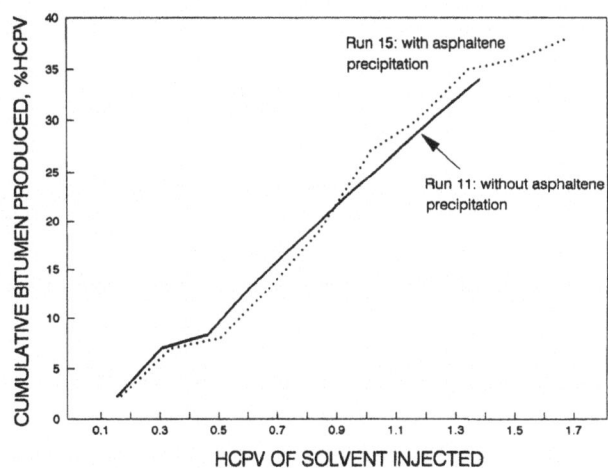

Figure 8. Effect of asphaltene precipitation on bitumen recovery for a thickness ratio of 12

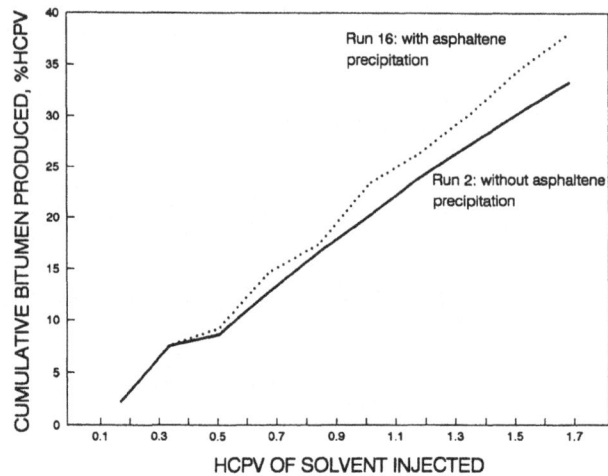

Figure 9. Effect of asphaltene precipitation on bitumen recovery for a thickness ratio of 2.25

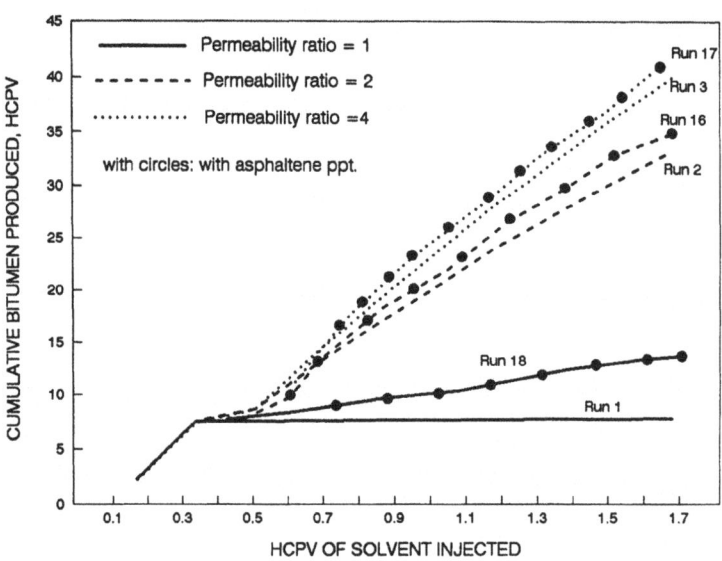

Figure 10. Effect of asphaltene precipitation on bitumen recovery for various permeability ratios

Following studies of solvent injection in absence of asphaltene precipitation, a series of numerical runs was conducted to model the effect of asphaltene precipitation. The first run was with a bitumen-to-water zone thickness ratio of 12. Other characteristics of the run (Run 15) are given in Table 1. Figure 8 compares results of this numerical run (Run 15) with those of Run 11 which was conducted without considering asphaltene precipitation. Note that the recovery is consistently lower during the early stages of solvent injection. During these early stages of recovery, asphaltene precipitation contributes to an increase in injection pressure through local plugging of the porous medium. While this effect is detrimental to bitumen production, it also helps decrease permeability of the bottom-water zone. The benefit of this phenomenon starts to show up during later stages of solvent injection. At the end of a solvent injection of 1 HCPV, the run with asphaltene precipitation recovers more bitumen than does the other run without asphaltene precipitation. This particular trend of higher ultimate recovery with asphaltene precipitation is even more intense when the oil-to-water zone thickness ratio is lower. Figure 9 compares results for Runs 2 and 16 for which the oil-to-water zone thickness ratio was 2.25. For this particular case the effect of asphaltene precipitation starts to show very quickly. A low thickness ratio leads to quicker invasion of the bottom-water zone which starts to plug up. Consequently, more solvent goes into the oil zone increasing bitumen leaching. Consequence of this phenomenon is manifested in higher bitumen recovery for the case which allows asphaltene precipitation. It appears from all these runs that the precipitation of asphaltene increases production by plugging the bottom-water zone. This effect is further shown in Figure 10. This figure shows recovery curves for a range of bitumen-to-water zone permeability ratios. Note that the difference between the asphaltene and non-asphaltene cases increases with decreasing permeability ratios. With all cases, however, asphaltene precipitation leads to some improvement over the case where such phenomenon is not considered.

CONCLUSIONS

A comprehensive numerical simulation approach was taken to model bitumen mobilization with solvents. Such process is important even when a thermal technique is being used. A new approach was taken to model asphaltene precipitation and its effect on bitumen recovery. The numerical simulation approach is tested on a bitumen reservoir containing a bottom-water zone. Such zone acts as a transporting medium for the solvent. It is shown that asphaltene precipitation improves bitumen recovery with solvent in all cases of permeability and thickness tested. However, the degree of improvement was higher when the bottom-water zone capacity was relatively higher. The series of parametric studies showed the presence of optimum flow rate as well as optimum bitumen-to-water zone thickness and permeability ratios.

REFERENCES

1. Moritis, G., "EOR Increases 24% Worldwide, Claims 10% of U.S. Production", Oil and Gas J., April 20, 1992, p. 260.

2. Murji, A., "Mobilization of Bitumen Under Reservoir Conditions", M.S. thesis, Dept. of Min. Met. Pet. Eng., University of Alberta, Edmonton, Canada, 1992, p.260

3. Alvarado, D.A., Ferrer, J.C., and Arteaga, A.E., "The Flow of Heavy Petroleum and Solvent Mixtures Through Porous Media", The Oil Sands of Canada -Venezuela, CIM Special Volume 17, 691-695, 1977.

4. Dealy, J.M., "Viscosity of Oil Sands Liquids", The Oil Sands of Canada-Venezuela, CIM Special Volume 17, 303-307, 1977.

5. Pirela, L. and Marcano, N.C., "Feasibility of Solvent Use in the Recovery of Heavy Crude Oils", The Oil Sands of Canada-Venezuela, CIM Special Volume 17, 307-309, 1977.

6. Cragoe, C.S., "Changes in the Viscosity of Liquids with Temperature, Pressure, and Composition", Proc. World Petroleum Congress, 1933, pp.529-543.

7. Chang, H.L., Farouq Ali, S.M., and George, A.E., "Peformance of Horizontal-Vertical Well Combinations for Steamflooding Bottom Water Formations", Annual Technical Meeting of the Petroleumociety of CIM, CIM/SPE 90-86, 1990.

8. Proctor, M.L., Georger, A.E., and Farouq Ali, S.M., "Steam Injection Strategies for Thin, Bottomwater Reservoirs", SPE paper no. 16338, presented at the California Regional Meeting, 1987.

9. Islam, M.R., "Recovering Oil from Bottom Water Reservoirs", Society of Petrolem Engineering, vol. 45; 514-516, June, 1993.

10. Islam, M.R. and Farouq Ali, S.M., "Numerical Simulation of Emulsion Flow Through Porous Media", paper 89-40-63 presented at the Annual Technical Meeting of the Petroleum Society of CIM, 1989.

11. Oguztoreli, M. and Farouq Ali, S.M., "A Mathematical Model for the Solvent Leaching of Tar Sand", SPE Reservoir Engineering, vol. 2; 545-555, Nov.1986.

12. Kamath, V.A., Islam, M.R., Patil, S.L., Jiang, J.C., and Kakade, M.G., "The Role of Asphaltene Aggregation in Viscosity Variation of Reservoir Hydrocarbons and in Miscible Processes", in *Particle Technology and Surface Phenomena in Minerals and Petroleum*, Eds. M. K. Sharma and G. D. Sharma, Plenum Press, 1991, pp.1-21.

NOMENCLATURE

C   concentration
C^* critical transition concentration
D   diffusivity, m^2/s
D_0 liquid phase molecular diffusivity, m^2/s
D_s solid phase molecular diffusivity, m^2/s
d_p grain particle diameter, m
F   convolution kernel
F_R formation resistivity factor
g   acceleration due to gravity, m/s^2
h   thickness, m
HCPV   hydrocarbon pore volume
h_b thickness of the bitumen zone, m
h_w thickness of the bottom-water zone, m
k   permeability, μm
k_b permeability of the bitumen zone, μm^2
k_w permeability of the bottom-water zone, μm_2
n_p particle number density
p   hydrodynamic pressure, kPa
r_p, r_o spherical particle radius, m
S   liquid phase saturation
t   time, s
v   Darcy velocity, m/s
u   mass flux, $kg/m^2/s$

Greek

α adsorption coefficient, m/s
β flow restriction parameter
δ transition layer thickness
μ viscosity, Pa.s
ϕ porosity
σ density, kg/m^3
σ packing inhomogeneity factor
τ dimensionless time for miscible displacement

ASPHALTENE CONVERSION DURING COAL-BITUMEN

CO-PROCESSING

A. Chakma

Dept. of Chemical and Petroleum Engineering
University of Calgary
Calgary, Alberta, T2N 1N4 (Canada)

ABSTRACT

Experimental studies on the conversion of asphaltene during the co-processing of coal and bitumen are described. Athabasca bitumen (15.4 % asphaltenes) was co-processed with a sub-bituminous coal (1.14% asphaltene) in a batch autoclave under hydrogen pressure at reaction temperatures varying from 400 to 440°C. Both thermal and catalytic co-processing experiments were conducted. The catalysts used were molten halide type and included $ZnCl_2$, $MoCl_5$, KCl, CuCl, and $SnCl_2$. Higher reaction temperature resulted in higher conversion of asphaltenes into both maltenes and coke and gases. As a result the H/C atomic ratio of the unconverted asphaltenes decreased with temperature. Higher reaction time on the other hand allowed maltenes to be converted to asphaltenes. While all the catalysts tested had catalytic effects on asphaltene conversion, $MoCl_5$ was found to provide the highest conversion of asphaltenes due to its ability to hydrogenate the radicals formed due to asphaltene cracking.

INTRODUCTION

Co-processing involves simultaneous upgrading of coal and bitumen/heavy oil in the presence of hydrogen into synthetic cruces. In co-processing coal is liquefied while bitumen/heavy oils are converted into lighter products. Co-processing is more advantageous than separate liquefaction of coal and upgrading of bitumen/heavy oil for many reasons. First of all, it eliminates or reduces the need for solvent recycle for coal liquefaction. Furthermore, co-processing

*Asphaltene Particles in Fossil Fuel Exploration, Recovery, Refining, and Production
Processes*, Edited by M.K. Sharma and T.F. Yen, Plenum Press, New York, 1994

47

provides higher liquid yields than what would be expected from linear addition of liquid yields obtained from separate coal liquefaction and bitumen/heavy oil upgrading operations. This is due to the fact that the bitumen/heavy oil fractions can act as hydrogen donors for the coal to some limited extent and the inorganic matters of the coal may provide some beneficial catalytic activity for the upgrading of bitumen/heavy oil. In addition, coal is also known to be capable of enhancing demetallation of the bitumen/heavy oil[1].

Extensive research work has been carried out on co-processing since the late seventies. High coal-oil conversion rates and a synergistic effect on the conversion have been reported by several authors[2,3]. The synergism is believed to be due to the solvolytic effects provided by the bitumen in which bitumen solubilizes the coal particles and therefore makes it easier to break its physical bonds. Bitumen may also act as a hydrogen donor solvent. The aliphatic hydrogen of the bitumen may help in the cleavage of coal structures. On its part, coal may enhance the upgrading of bitumen due to the presence of inorganics such as pyrite, which may provide beneficial catalytic effects for the upgrading of bitumen. An extensive demetallation, deoxygenation and desulphurization of the liquid products have also been reported[4].

Fouda et al.[5] investigated the effect of coal concentration on the co-processing performance and observed a marked improvement in the process operability with the addition of coal, as compared to the processing of Cold Lake vacuum bottoms only. They found the extent of oxygen, vanadium and nickel removal to increase while coke formation was found to decrease. It is believed that metals such as Vanadium, are adsorbed onto the surface of the unconverted coal.

While thermal co-processing provides all the previously mentioned advantages of co-processing, the liquid yields are rather low and require operation under high severity conditions. As such catalytic co-processing is the preferred choice. The use of a catalyst may enhance coal conversion and liquid yields. Several different catalysts have been investigated for co-processing.

Molten halide catalysts have long been used in the liquefaction of coal. Zielke et al.[6] found molten $ZnCl_2$ to be superior over conventional catalysts for hydrocracking of pyrene, coal and coal extracts. They also carried out continuous hydroliquefaction of sub-bituminous coal over molten $ZnCl_2$ and obtained octane rich gasoline-range products[7]. The catalytic action is suggested to proceed through an ionic mechanism that cleaves bonds in polyaromatic compounds, but is unable to open mono-aromatic rings[8].

Nomura and coworkers[9-14] through a series of studies have shown good conversion rates as well as high liquid yields from different coals, when combinations of several different metal chlorides were used as catalysts. Combination of $ZnCl_2$ and one or several other metal chlorides ($MoCl_5$, KCl, CuCl, $CrCl_3$) were shown to be more active than $ZnCl_2$ alone, the $ZnCl_2$ - $MoCl_5$ mixture being the most effective. Also, the amount of consumed hydrogen, was lowest when this mixture was used.[10] $SnCl_2$ containing salts were shown to give higher yields of both hexane-soluble and benzene-soluble fractions, and lower yields of gases, than $ZnCl_2$, and $ZnCl_2$-KCl-NaCl salts[12]. $ZnCl_2$ has also been found to be an excellent catalyst for converting asphaltenes into maltenes.

Nomura et al.[9] found that an addition of $MoCl_5$ to a mixture of $ZnCl_2$-KCl-NaCl gave higher conversion of asphaltenes to maltenes and a markedly reduced coke formation (3.7% compared to 15.6 %), suggesting very good hydrocracking abilities of the $MoCl_5$ on asphaltenes. Also the sulfur contents of the pentane-soluble and benzene-soluble fractions were lower when $MoCl_5$ containing catalysts were used.

Herrman et al.[15] have compared the performance of Fe, $ZnCl_2$ and $ZnCl_2$-promoted Fe catalysts for the hydrocracking of Athabasca bitumen. They found $ZnCl_2$ to be the most active and to produce less hydrocarbon gas and significantly less sulphur in the liquid product. Chakma et al.[16] treated Athabasca coker feed bitumen with $ZnCl_2$, CuCl, and mixtures of the two, under a continuous flow of hydrogen and found that $ZnCl_2$ was effective for the conversion of asphaltenes to maltenes and for lowering the coke formation. CuCl was not as effective, and decreased the activity of $ZnCl_2$ when added to it.

OBJECTIVES

The primary objectives of the present work are:

1. To obtain a better understanding of the fate of asphaltenes in the co-processing reactions under catalytic and non catalytic conditions, and

2. To understand the interactions between asphaltene and other fractions of heavy oils during co-processing.

To achieve the above stated objectives an experimental program was initiated. The experimental program mainly consisted of reacting bitumen, coal-bitumen mixtures, and asphaltene-solvent mixtures with and without the presence of catalysts under H_2 pressure (typically 6.9 MPa) at desired temperatures for 1 to 2 hr periods and analyzing of the reaction products.

EXPERIMENTAL

Reagents

Athabasca coke feed bitumen, produced by the hot water process, was supplied by the Alberta Research Council sample bank. It was used without any further preparations. Select properties are shown in Table I.

TABLE I. SELECT PROPERTIES OF ATHABASCA COKER FEED BITUMEN

PARAMETERS	DATA
API gravity:	9.0
Density (kg/m^3)	1006
Viscosity 25°C (mPa.s)	197,000
Ash (wt%)	0.70
Carbon (wt%)	83.16
Hydrogen (wt%)	10.27
Nitrogen (wt%)	0.47
Sulphur (wt%)	4.95
Oxygen (wt%)	0.76
Asphaltenes (wt%)	15.4
H/C-ratio	1.47

Sub-bituminous coal was supplied by Forestburg Colliers Ltd. It was pulverized and sieved to pass through <75 mm screen (200 Tyler Mesh) . Select properties of the coal used are shown in Table II.

TABLE II. SELECT PROPERTIES OF SUB-BITUMINOUS COAL

PARAMETERS	DATA
Ash (wt%)	19.9
Carbon (wt%)	67.59
Hydrogen (wt%)	4.28
Nitrogen (wt%)	1.48
Residue (wt%)	26.35
H/C-ratio	0.76

$ZnCl_2$, $MoCl_5$, KCl, CuCl, NaCl, $SnCl_2$ (certified ACS grade), were supplied by Fischer Scientific Company. Reagent grade n-pentane and toluene were provided by Anachemia Science. Omnisolv methanol with a minimum purity of 99.9% was supplied by BDH Inc.

Hydrogen (g) and nitrogen (g), with purities exceeding 99.95% were supplied by Medigas Canada Inc.

Apparatus

The experiments were carried out in a stainless steel, bolted closure, stirred autoclave with an internal volume of 300 mL. Agitation was achieved by a magnetically actuated, packless and non-contaminating rotary impeller system. The impeller was designed to draw the process gases (hydrogen) through the tubular drive shaft and disperse it through the liquid phase at high velocity. The agitator was driven by a DC motor at 1800 rpm.

The autoclave was heated by an electric, external jacket type heater, which was controlled with a PID temperature controller.

Impregnation of Coal with Molten Catalysts

Impregnation of coal with molten catalysts was carried out by soaking the coal with a catalyst saturated methanol solution. The catalyst and coal were dried for 2 hr. at 105°C. Known amount of dry catalyst was then dissolved in

200 mL of methanol. Appropriate quantity of dry coal was then added to the mixture. The mixture was continuously stirred with a magnetic stirrer. After soaking the coal overnight, the mixture was filtered through a coarse fitted disc funnel and a Whatman # 2 filter paper. The funnel with the methanol/catalyst soaked coal was then dried for up to 48 hr. or to constant weight, in a vacuum oven at 75°C. The amount of catalyst impregnated can then be determined.

Procedures

The autoclave was cleaned before each run with acetone followed by dichloromethane and tetrahydrofuran to remove any catalyst and organics that might have been left from the previous experiment.

40-45 g of the catalyst impregnated coal was dried for 2 hr. at 105°C. After cooling to room temperature in an evacuated desiccator, the total weight of the dry coal, catalyst and the container (beaker) was recorded. The coal was then transferred into the autoclave, and the total amount of coal transferred was found from material balance. Bitumen was then added to the autoclave. The ratio of bitumen to coal was maintained at 70:30.

The autoclave was then sealed and pressure tested with nitrogen. After it had been determined that no leaks were present, the pressure was released very slowly to prevent bitumen and/or coal from being trapped and blown out of the reactor. To remove as much atmospheric oxygen and moisture as possible, the autoclave was purged once more.

The autoclave was then pressurized with hydrogen to an initial pressure of 6.9 MPa at ambient temperature (22°C), and heating of the reactor was started. After heating to the desired temperature, the temperature was held constant for either 1 or 2 hr. At the end of the experiment, cold water was circulated through an internal cooling coil. The temperature could be brought down to 100 °C in about 30 min., and to room temperature in about an hour. The heating and cooling rates as well as the temperature fluctuations were monitored and recorded.

After cooling the autoclave to room temperature, the pressure was carefully released. When the autoclave had reached atmospheric pressure, all connections were removed, and the reactor was cleaned for product recovery.

Product Workup Procedure

The reaction products were separated into different fractions according to Figure 1. The total product was removed from the autoclave with toluene. The walls of the reactor as well as all tubing and other parts were carefully scraped to remove as much coke and catalyst as possible. Larger aggregates or lumps of coke and catalyst were crushed with a mortar and pestle to prevent toluene soluble material from being trapped within such lumps.

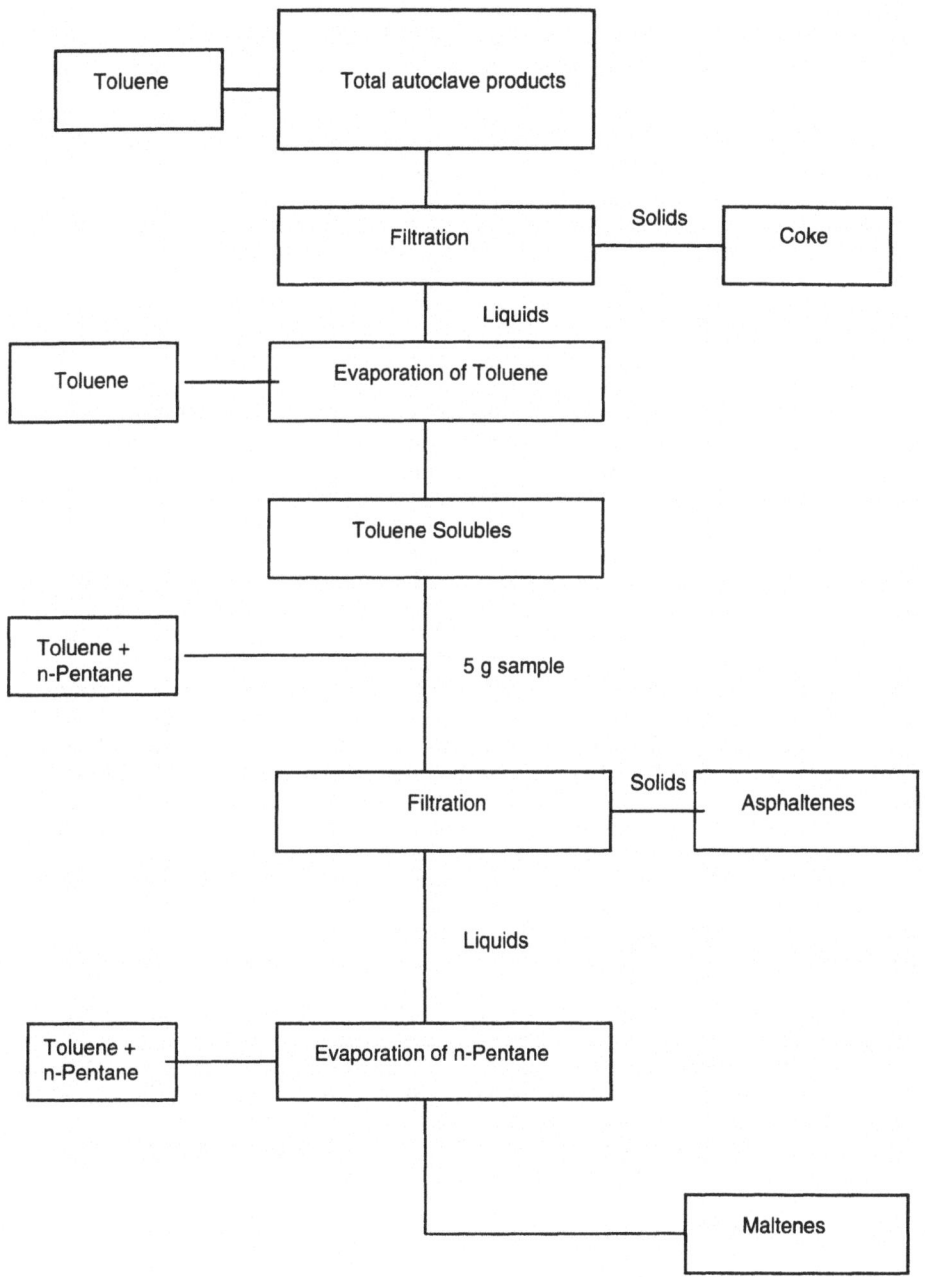

Figure 1. Basic product workup procedure.

To ensure that all toluene soluble material was dissolved, the products together with 0.8-1.0 L toluene were placed in the ultrasonic bath for 5 min. at room temperature. The mixture was then filtered through a weighed Whatman # 2 filter and a weighed coarse (ASTM 40-60) fitted disc funnel. This separated the toluene insolubles (TI). The filter cake was washed repeatedly with a total of approximately 0.8 L toluene, so that the total volume of toluene used was approximately 1.6 L.

The fitted disc funnel, with the filter cake, was dried to constant weight in a vacuum oven at 75°C. The weight of the separated solids was the yield of the TI fraction, which contained coke, unreacted coal, catalyst and ash. All of the $ZnCl_2$ and ash included in the feed was considered to be recovered in the TI fraction.

The filtrate was transferred to a tarred round bottom flask. The toluene was evaporated in the Rotavapor at 65°C and 90 rpm. The pressure of the rotavapor was gradually reduced to 50 mbar, and the end point of the evaporation was chosen to be when no more condensation could be observed at this pressure. The weight of the remaining liquid product was the yield of the toluene soluble fraction (TS).

The toluene solubles (TS), were further fractionated into asphaltenes and maltenes (distillable, pentane soluble oils), according to a procedure proposed by Moschopedis and Hawkins,[17] with some minor modifications. 5 g of the TS was dissolved in 5 g toluene, in an Erlenmeyer flask in an ultrasonic bath, the asphaltenes were filtered off through a Whatman # 2 filter paper in a tarred medium porosity (ASTM 10-15) fitted disc funnel. The filter cake was crushed and rinsed with an additional 200 mL pentane. Any asphaltenes left on the walls of the Erlenmeyer flask were dissolved in a small amount of toluene and transferred to the fitted disc funnel. The funnel was dried in the vacuum oven at 75°C until constant weight, and then cooled to room temperature in a desiccator. The weight of the filter cake was the weight of the pentane insoluble fraction (PI), or asphaltenes.

The pentane was evaporated in the Rotavapor at 65°C and 830 mbar. The pressure was then lowered to 40 mbar to evaporate the remaining toluene. The evaporation was stopped when the weight of the remaining pentane solubles (PS), also called maltenes, and the weight of the PI, or the asphaltenes, approximately equaled the weight of the original sample taken from the TS fraction.

Analyses of Products

The viscosity of the total toluene soluble fraction was measured, when sufficient volume of this fraction was recovered. This was done using a Contraves Rheomat 115 rotational viscometer with a MS-DIN 125 measuring system.

The coke and the maltene fractions were analyzed for C, H and N content using a Control Equipment Corporation Model 240-MA elemental analyzer.

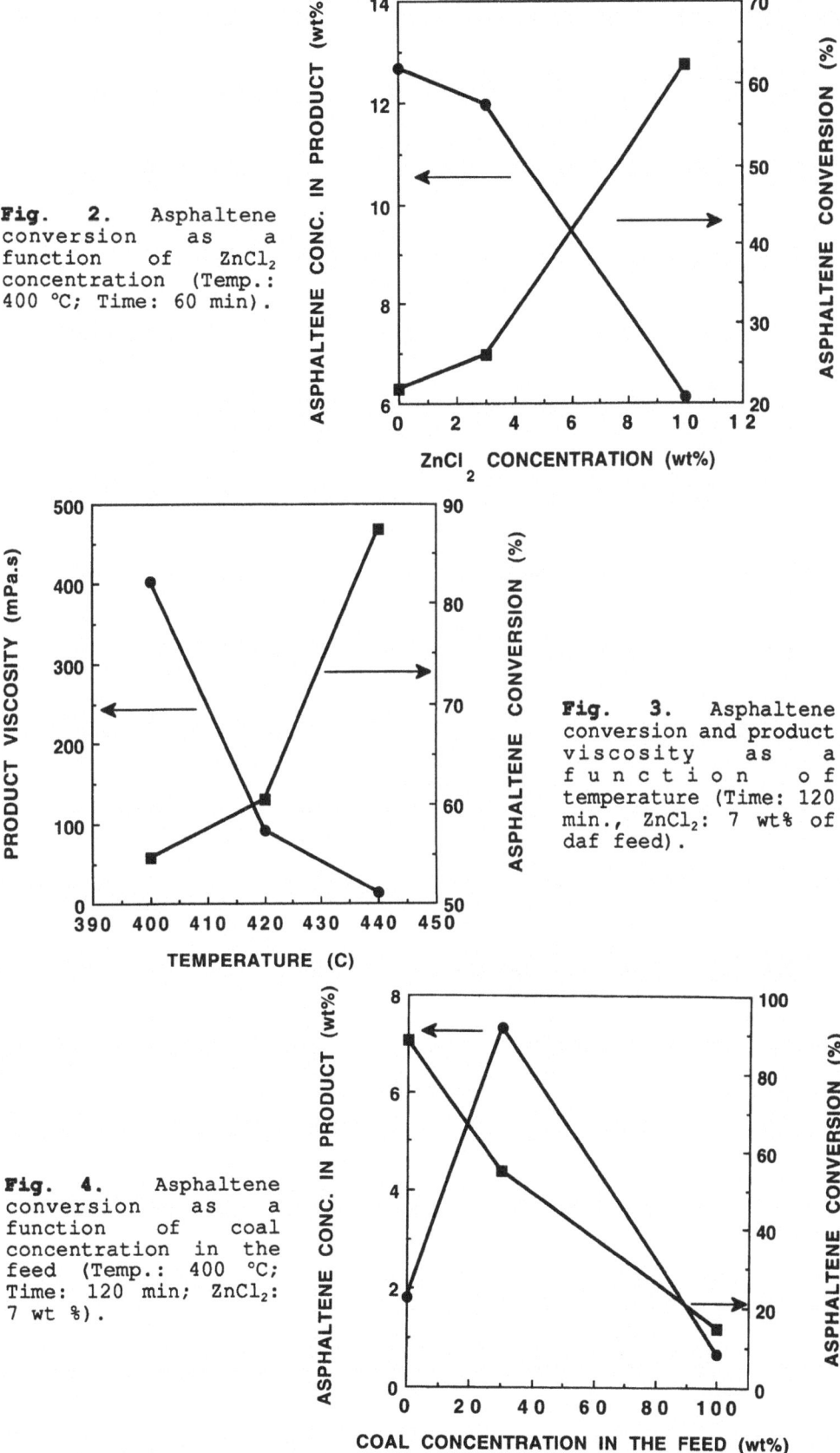

Fig. 2. Asphaltene conversion as a function of $ZnCl_2$ concentration (Temp.: 400 °C; Time: 60 min).

Fig. 3. Asphaltene conversion and product viscosity as a function of temperature (Time: 120 min., $ZnCl_2$: 7 wt% of daf feed).

Fig. 4. Asphaltene conversion as a function of coal concentration in the feed (Temp.: 400 °C; Time: 120 min; $ZnCl_2$: 7 wt %).

RESULTS AND DISCUSSION

Experiments with ZnCl$_2$ Catalysts

Figure 2 shows asphaltene concentration in the product stream and asphaltene conversion*, as a function of ZnCl$_2$ concentration for experiments carried out at 400 °C, under H$_2$ partial pressure of 6.9 MPa, for two hours. Coal/bitumen wt. ratio was 30:70. Asphaltene conversion without any catalyst was slightly over 20%. It increased with ZnCl$_2$ concentration and was over 63% with 10 wt% of ZnCl$_2$. It is clear that ZnCl$_2$ promotes asphaltene concentration significantly.

Figure 3 shows the effect of reaction temperature on product viscosity and asphaltene conversion with 7 % ZnCl$_2$ catalyst. As the temperature rises asphaltene conversion also increases. This is accompanied by a significant reduction in the liquid product viscosity. While the decreasing asphaltene content is not solely responsible for the viscosity reduction, it certainly plays a very important part.

Figure 4 shows the effect of coal concentration in the feed on asphaltene conversion and asphaltene concentration in the product. Increasing the coal content in the feed results in an initial increase and subsequent decrease of asphaltene in the product. This trend could be misleading. The main reason for the initial increase is the decline in the amount of solvent available for the coal-bitumen mixture as compared with pure bitumen. The lower asphaltene concentration in the product stream for pure coal is simply due to the lower asphaltene content of coal. Asphaltene conversion provides a better idea on the effect of coal concentration in the feed. With pure bitumen and 7 wt% ZnCl$_2$ as catalyst, asphaltene conversion is over 84%. With pure coal, the conversion is less than 9%. With 30% coal and 70% bitumen, it is about 54%. This is understandable. Pure bitumen provides higher solvolytic effects than the mixture, while pure coal does not provide and solvolytic effects in the absence of solvent recycle.

Figure 5 shows the effect of reaction time on asphaltene conversion with ZnCl$_2$ as a catalyst. Initially, conversion increases with reaction time, reaches a maximum and decreases later. The exact location of this maximum can not be found from Figure 5 due to the limited number of data points. Beyond this maximum, cracked maltenes start to combine and form asphaltenes, thereby setting up a competitive process where asphaltenes are cracked while maltenes recombine to form asphaltenes and the latter dominates.

Experiments with Mixed Catalysts

Figure 6 shows a comparison of 10% ZnCl$_2$, 12% MoCl$_5$ and a combination of 5.25% ZnCl$_2$-1.75% MoCl$_5$ catalysts in terms

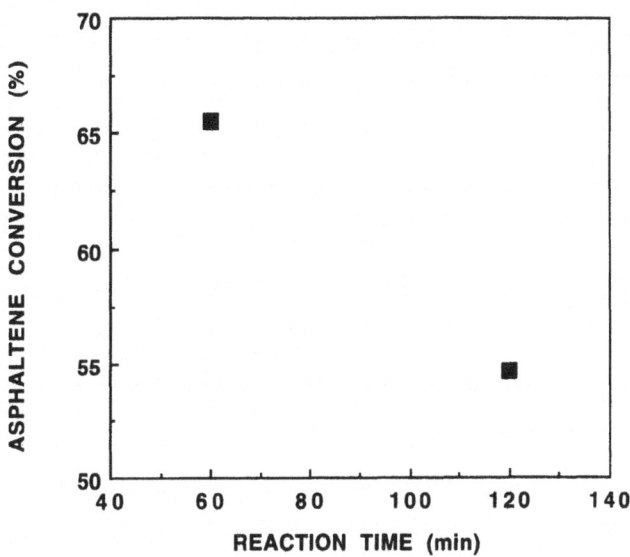

Figure 5. Asphaltene conversion as a function of reaction time (Temp.: 400 °C; $ZnCl_2$: 7 wt% daf feed).

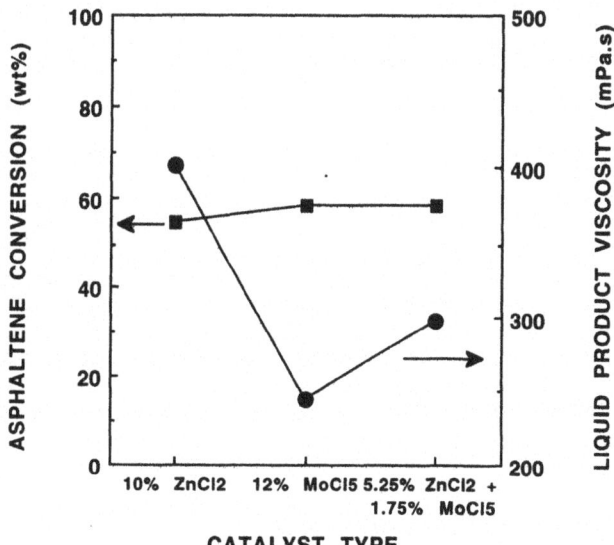

Figure 6. Asphaltene conversion and liquid product viscosity for different catalysts (Temp.: 400 °C; Time: 120 min.).

of asphaltene conversion and corresponding liquid product viscosity. Among the three catalyst formulations, $ZnCl_2$ provided the lowest asphaltene conversion. Catalyst formulations with $MoCl_5$ provide slightly higher conversions. However, the liquid product viscosities are significantly different for $ZnCl_2$ and $MoCl_5$ based catalysts. Liquid product obtained with $MoCl_5$ has the lowest product viscosity. This is due to the better hydrogenation activity of $MoCl_5$ compared with $ZnCl_2$ as can be seen from the H/C ratio of the maltenes and asphaltenes shown in Figure 7. The atomic H/C ratio of the maltenes, which constitute the bulk of the liquid product is higher for $MoCl_5$ based catalysts than $ZnCl_2$.

The liquid product viscosity and asphaltene conversion as a function of catalyst combination is shown in Figure 8. The asphaltene conversion remains almost the same with catalyst combinations studied, while liquid product viscosity decreased significantly by changing catalyst combination from $ZnCl_2$-$MoCl_5$ to $ZnCl_2$-$SnCl_5$.

The effect of temperature on asphaltene conversion and H/C ratio with $ZnCl_2$-$MoCl_5$-KCl mixtures can be seen from Figure 9. While asphaltene conversion increases with temperature, H/C ratio of the liquid product decreases, primarily due to the conversion of part of the liquid product into gases.

A number of experiments were carried out to determine the effect of reaction temperature and time on co-processing with the optimum mixed catalyst composition. The reaction temperatures for these runs were fixed at 400°C. The catalyst was 7 wt% $ZnCl_2$-$MoCl_5$-KCl (3:1:1). The reactor was pressurized with H_2 to a pressure of 6.89 MPa.

Figure 9 shows the effect of reaction temperature on asphaltene conversion and H/C ratio. As with the $ZnCl_2$ catalyst, asphaltene conversion increases with temperature, while the H/C ratio of the asphaltenes decreases.

Figure 10 shows the effect of reaction time on asphaltene conversion and H/C ratio of the liquid product with $ZnCl_2$-$MoCl_5$-KCl mixtures. Initially, asphaltene conversion increases with reaction time but decreases later. This is due to the conversion of some of the maltenes into asphaltenes. Also at higher reaction time more gas is formed which results in the decrease of the H/C ratio.

Asphaltene - Solvent Interactions

Experiments were carried out by reacting asphaltenes separated from bitumen and different solvents without any catalysts. The solvents used were: de-asphalted and de-coked bitumen (labeled as bitumen in the figure), maltene fraction obtained from co-processing of coal and bitumen, a hydrogen donor solvent, tetralin, and decalin. Temperature for all these runs were kept at 400 °C and the reaction time was 1 hr. Asphaltene to solvent ratio was 1:4.3.

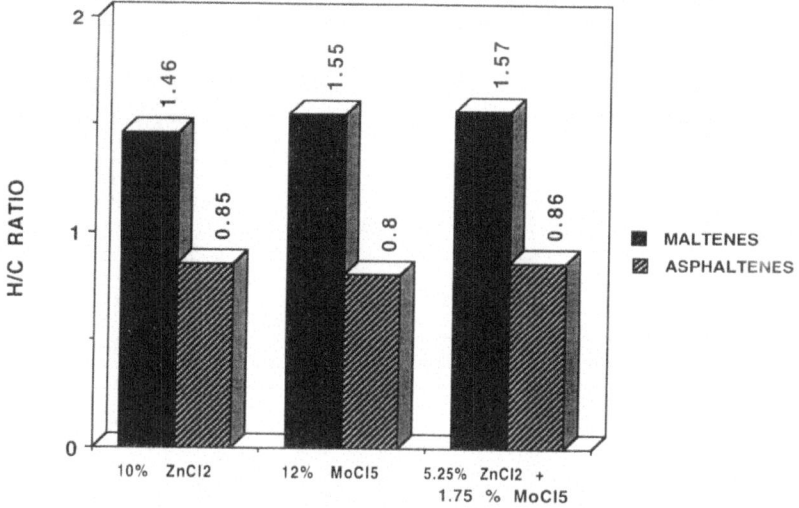

Figure 7. H/C ratio of maltenes and asphaltenes for different catalysts (Temp.: 400 °C; Time: 120 min.).

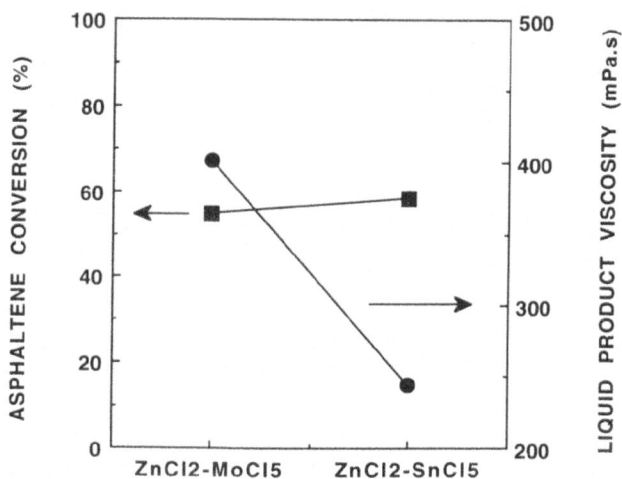

Figure 8. Asphaltene conversion and liquid product viscosity for different catalyst combination (Temp.: 400 °C; Time: 120 min.).

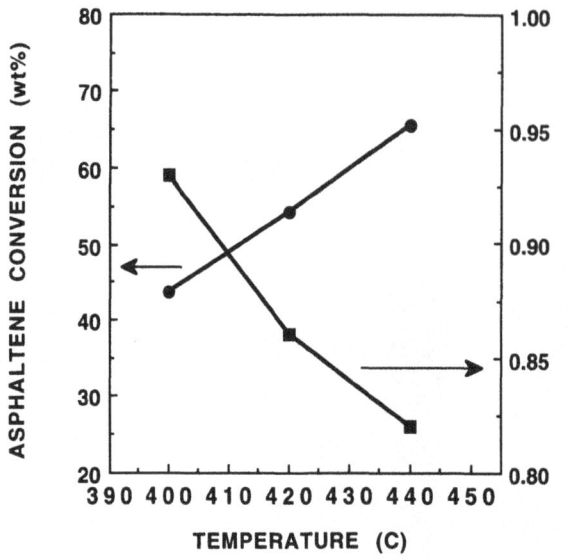

Fig. 9. Asphaltene conversion and H/C ratio as a function of temperature for the mixed catalyst combination of $ZnCl_2$-$MoCl_5$-KCl (Temp.: 400 °C; Time: 120 min.).

Figure 11 shows product yield for the non-catalytic case. As can be seen, processing of asphaltene with tetralin provided the lowest yield of asphaltene, thus indicating the highest asphaltene conversion. It also provided the highest yield of maltenes. This superior performance of tetralin can be attributed to its hydrogen donor capability. The worst performance was obtained with decalin. The de-asphalted and de-coked bitumen was superior to decalin. The performance of the maltene fraction obtained from co-processed liquid was superior to that of de-asphalted and de-coked bitumen, but inferior to that of tetralin. This proves that maltene obtained from co-processed liquid has better hydrogen donor properties than de-asphalted bitumen.

Fig. 10. Asphaltene conversion and H/C ratio as a function of reaction time for the mixed catalyst combination of $ZnCl_2$-$MoCl_5$-KCl (Temp.: 400 °C).

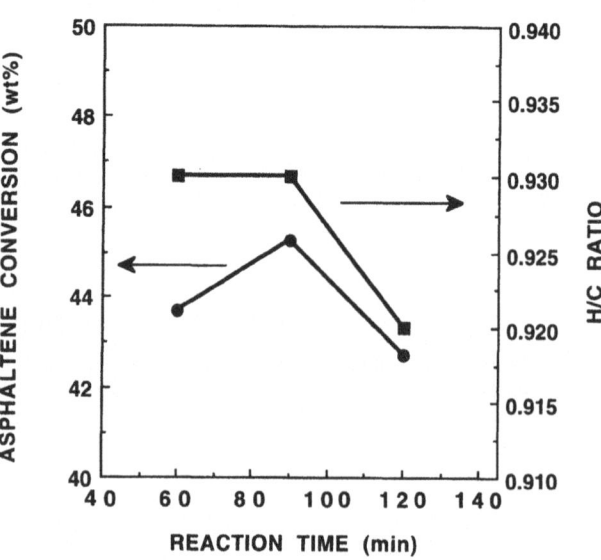

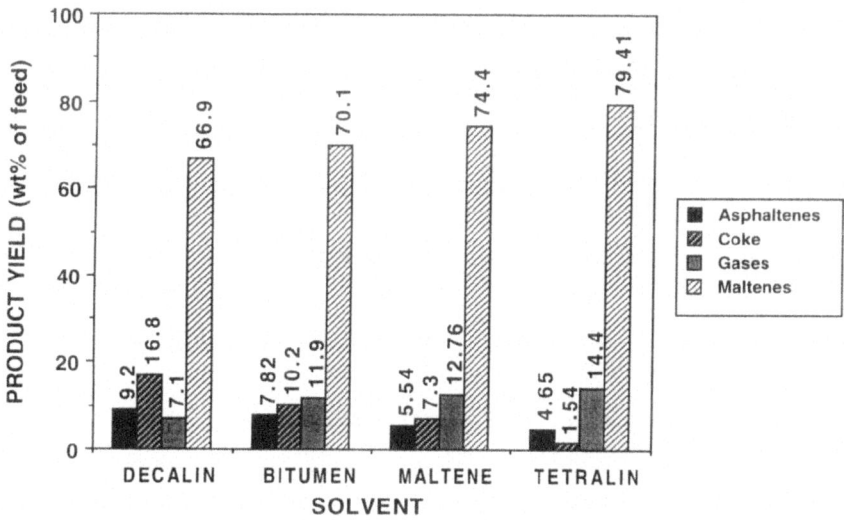

Figure 11. Product yield when asphaltene-solvent mixtures were cracked under hydrogen pressure at 400 °C for 1 hr.

CONCLUSIONS

Molten halide catalysts such as $ZnCl_2$, $MoCl_5$, KCl, CuCl, and $SnCl_2$ are effective in cracking asphaltenes during coal-bitumen co-processing. Higher reaction temperature results in higher conversion of asphaltenes into both maltenes and coke and gases. As a result the H/C atomic ratio of the unconverted asphaltenes decreases with temperature. Higher reaction time on the other hand allows maltenes to be converted to asphaltenes. While all the catalysts tested had catalytic effects on asphaltene conversion, $MoCl_5$ was found to provide the highest conversion of asphaltenes due to its ability to hydrogenate the radicals formed due to asphaltene cracking. Tetralin was found to be a good hydrogen donor solvent for asphaltene cracking. Maltene fractions from co-processed liquids also act as good solvents.

ACKNOWLEDGEMENTS

The financial assistance provided by the Office of Coal Research and Technology of Alberta Energy of the Province of Alberta and the Natural Sciences and Engineering Research Council of Canada (NSERC) in support of this work are gratefully acknowledged. Mr. Y. Hu and C.J. Franzen assisted in the laboratory work.

REFERENCES

1. Lett, R.G. and Cugini, A.V. Proceedings of the DOE Direct Liquefaction Contractors' Review Meeting, October 22, 1986.

2. Boehm, F.G., Caron, R.D., Banarjee, D.K. Energy and Fuels, **3**, 116, 1989.

3. Lenz, U. , Wawrzinek, J. and Giehr, A. Preprints of ACS Division of Fuel Chemistry, **33**(1), 27, 1988.

4. Cugini, A.V., Lett, R.G. and Wender, I. Energy and Fuels, **3**, 12, 1989,.

5. Fouda, S.A., Kelly, J.F. and Rahimi, P.M., Energy and Fuels, **3**, 154, 1989,

6. Zielke, C.W., Struck, R.T., Evans, J.M., Costanza, C.P. and Gorin, E. Ind. Eng. Chem. Process Des. Dev. 5, 151, 1966,

7. Zielke, C.W., Klunder, E.B., Maskew, J.T. and Struck, R.T., Ind. Eng. Chem. Process Des. Dev., **19**, 85, 1980.

8. Derbyshire, F. Energy and Fuels, **3**, 273, 1989.

9. Nomura, M., Terao, K., and Kikkawa, S., Fuel, **60**, 699, 1981.

10. Nomura, M., Miyake, K., Kikkawa, S., Fuel, **61**, 18,1982,

11. Nomura, M., Kimura, K., Kikkawa, S., Fuel, **61**, 1119, 1982,

12. Nomura, M., Sakashita, H., Miyake, M. and Kikkawa, S., Fuel, **62**, 73, 1983.

13. Nomura, M., Kawakami, Y. and Miyake, M., Fuel Process Technol, **14**, 261, 1986.

14. Song, C., Nomura, M., and Miyake, M., Fuel, **65**, 922, 1986.

15. Herrmann, W.A.O., Mysak, L.P. and Belinko, K., CANMET Report 77-50, Energy, Mines and Resources Canada, Ottawa, 1977.

16. Chakma, A., Chornet, E., Overend, R.P. and Dawson, W.H., Energy and Fuels, **3**, 144, 1989.

17. Moschopedis, S.E. and Hawkins, R.W., Alberta Research Council, Information Series 94, 1981.

CLASSIFICATION OF ASPHALT TYPES

BY ASPHALTENE AROMATICITY

H. J. Lian and T. F. Yen

Environmental and Civil Engineering
University of Southern California
Los Angeles, CA 90089-2531

ABSTRACT

A new method is presented using the correlation of ^{13}C NMR aromaticity vs. H/C of the derived asphaltene for asphalt. The parent asphalt can clearly be separated into three types. This means that the component has to be separated from the parent asphalt first before the correlation. Result from this agrees with the traditional penetration index classification. In this paper, correlation between asphaltene aromaticity and temperature for different sources of asphalts or petroleum has been discussed.

INTRODUCTION

Asphalt is a very complex mixture which is obtained primarily from bottom materials of the petroleum refining process. It consists of saturates, aromatics, resins, and asphaltenes (Fig.1). These four fractions can be separated by column chromatography (i.e. ASTM D4124[1], solvent fractionation[2] or SARA[3]), planar chromatography (i.e. thin-layer Chromatotron[4], or thin-layer Chromarod (TLC-FID)[5]).

Due to its special properties such as impermeability to water, high adhesiveness and cohesiveness, susceptibility to temperature change, resistance to abrasion, acid, corrosion

Asphaltene Particles in Fossil Fuel Exploration, Recovery, Refining, and Production
Processes, Edited by M.K. Sharma and T.F. Yen, Plenum Press, New York, 1994

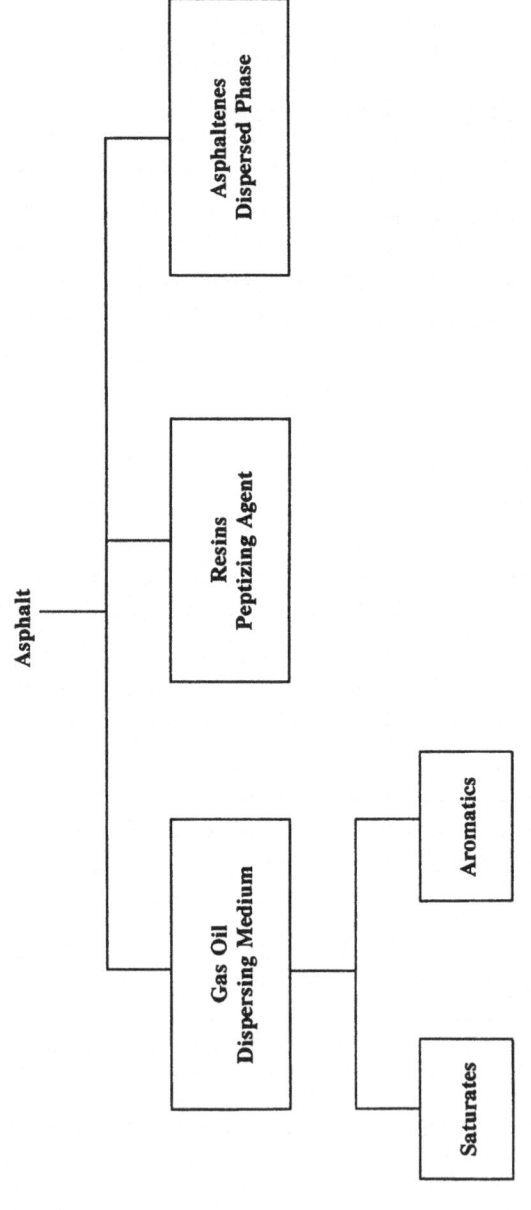

Figure 1. The basic four components of asphalt

Table I. Rheological characteristics of three colloidal types of asphalt

Rheology property	Sol	Sol-Gel	Gel
	Solvent	Vacuum or Steam	Air-blown
	Newtonian	Viscoelastic	Elastic
Complex flow, c	1.0	0.85-1.0	0.4-0.85
Asphalt aging	<0.02	0.02-0.09	>0.1
Plastic flow, p	1.0	1.0-1.4	>1.4
Penetration index	<-1.0	-1.0-1.0	>1.0
Ductility at 25 °C	very high	high (approx.)	very low to none
C/H ratio (atomic) Asphaltenes	high (0.95-1.3)	low (0.65-0.90)	low (0.6-0.97) or high (0.9-1.3)
Malthenes	high (0.95-1.3)	about 0.7	low relative to the above ratio

environment, etc., asphalt can be used in paving, roofing, coating, sealing, etc.[6] Because of different contents in various sources of asphalts, their colloidal properties are quite different. Therefore asphalt can be applied to different fields.

In general, asphalts can be classified into sol, sol-gel, and gel types depending on their physical and chemical properties. Sol and sol-gel type asphalts can be used on paving, while gel type asphalts are applied in roofing. The majority of air-blown asphalts belong to the gel type. Table I lists some of the traditional indexes such as complex flow c, asphalt aging index, plastic flow p, penetration index, ductility, and the C/H atomic ratio of asphaltenes which can be used to distinguish three different types of asphalt[7]. The most popular method, which has been adopted by industry, is the penetration index because it makes it easy to determine by Equation 1, if one has penetration and softening point data and if the current gradation of asphalt in industry is using viscosity or penetration. Those penetration and softening point values are commonly tested in industry by ASTM D36-86 (softening point, ring & ball method) and ASTM D5-86 (penetration test). Figure 2 indicates that the penetration indexes of all blown asphalts are higher than 1 which coincide with the range of penetration indexes for gel type asphalt in Table I[8]. Table II lists the penetration and softening point data of six core asphalts which were supported by the Strategic Highway Research Program (SHRP)[9]. They can easily be classified into three different types of asphalt by using Equation 1 to figure out each penetration index. From Table I, AAG-1 can be classified as a sol type, AAA-1, AAB-1, AAK-1, AAM-1 are sol-gel types, and AAD-1 belongs to a gel type. The sources and refinery processes of six core asphalts are listed on Table III.

Penetration index (P.I.):

$$\frac{20-P.I.}{10+P.I.} \times \frac{1}{50} = \frac{\log 800 - \log \textit{penetration at } T^\circ C.}{\textit{softening point } (R \wedge B,^\circ C.) - T^\circ C.} \qquad (1)$$

For this paper, we attempt to use asphaltene aromaticity measured by nuclear magnetic resonance (NMR) with the H/C ratio of asphaltene to distinguish different types of asphalt. The relationship between temperature and asphaltene aromaticity will also be discussed.

THEORY AND EXPERIMENTAL

THEORY: A Bruker AC250 Pulsed Fourier Transform NMR Spectrometer, with a continuous-wave circuit, a computer-controllable pulse generator, and a digital computer (HP7550) was used in this experiment. The asphaltene solution was subjected to a strong magnetic field generated by a magnet and transmitter. Each spinning nucleus absorbed the energy to transit from one alignment in the applied

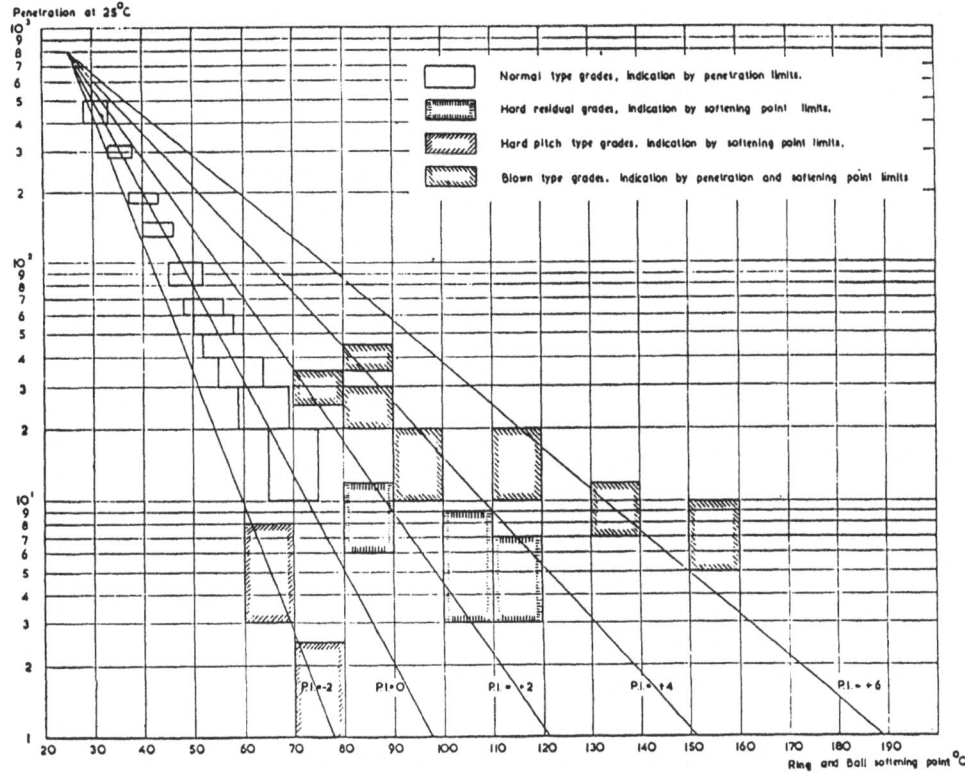

Figure 2. Scheme of available bitumen grades

Table II. Identification of six core asphalts by penetration index

	Penetration (100 g, 5s): 25 °C, 0.1 mm	Softening point (R&B), °C	Penetration index (P.I.)	Colloidal Type
AAA-1	160	44.4	0.72	sol-gel
AAB-1	98	47.8	-0.01	sol-gel
AAD-1	135	47.8	1.13	gel
AAG-1	53	48.9	-1.35	sol
AAK-1	70	49.4	-0.52	sol-gel
AAM-1	64	51.7	-0.19	sol-gel

field to another one. The amount of energy required to cause a particular nucleus to realign depends on the field strength, the electronic configuration around the particular nucleus, the anisotropy, the type of molecule, and the intermolecular interactions[10]. From the characteristic absorption of energy by certain spinning nuclei, we can identify atomic configurations in molecules. The absorption energy through the RF oscillator will send the signal to the computer after it has been amplified and the interference filtered out. The results are accurate, especially for the dilute ^{13}C nuclear spin. Finally, the aromatic to aliphatic compounds ratio can be measured by comparing the height of two different regions which were drawn by the printer.

EXPERIMENTAL: First, six different asphaltene samples were isolated from six core asphalts (Table III) by n-pentane. Then each asphaltene sample was dissolved in a 99.8% chloroform solvent at a 10-15% concentration range and placed into 30 cm glass tubes respectively. Proton NMR ^{13}C spectra were obtained from a Bruker AC250 spectrometer. The line of aliphatics and aromatics was drawn with a HP7550 computer. Aromaticity was easily determined by measuring the peak height of aliphatics (H_{al}) and the peak height of aromatics (H_{ar}). Lastly, aromaticity of each asphaltene was obtained by using the following formula:

$$Aromaticity = H_{ar} / (H_{ar} + H_{al}) \qquad (2)$$

Next, 4.5 grams of AAG-1 asphalt were kept in the crucible covered by aluminum foil, and heated in a high temperature oven at 156, 230, and 340 °C separately where 1.5 grams samples were taken out at each temperature after the sample was cool enough. Those three samples in addition to the 1.5 gram original sample were dissolved respectively in 10 mL of chloroform in separate 30 mL beakers. By following the above procedures each sample's NMR spectrum was obtained.

RESULTS AND DISCUSSIONS

Figures 3 to 8 are the NMR spectra of six asphaltenes which have negative linear correlationships in aromaticity vs. H/C ratios. H/C ratio data originated from Western Research Institute (WRI). Aromaticity and H/C ratio data are printed in Table IV. Figure 9 shows that AAG-1 and AAM-1 have a -1 slope; AAB-1, AAA-1, AAK-1 have a -1.38 slope; and AAD-1 has no relationship with the other samples. Evidently, the lower the negative slope is, the more the asphaltene is likely to be a sol type. AAG-1 and AAM-1 could be sol types; AAA-1, AAB-1, and AAK-1 could be sol-gel types; and AAD-1 belongs to another group: the gel type. The results are quite similar to the penetration index with the exception of AAM-1. More evidence, however, shows that AAM-1 is a sol type asphalt. For instance, from Table III, AAM-1 core asphalt comes from a solvent distillation process. Also, the asphaltene content of AAM-1 is not high enough, being about 4%[9].

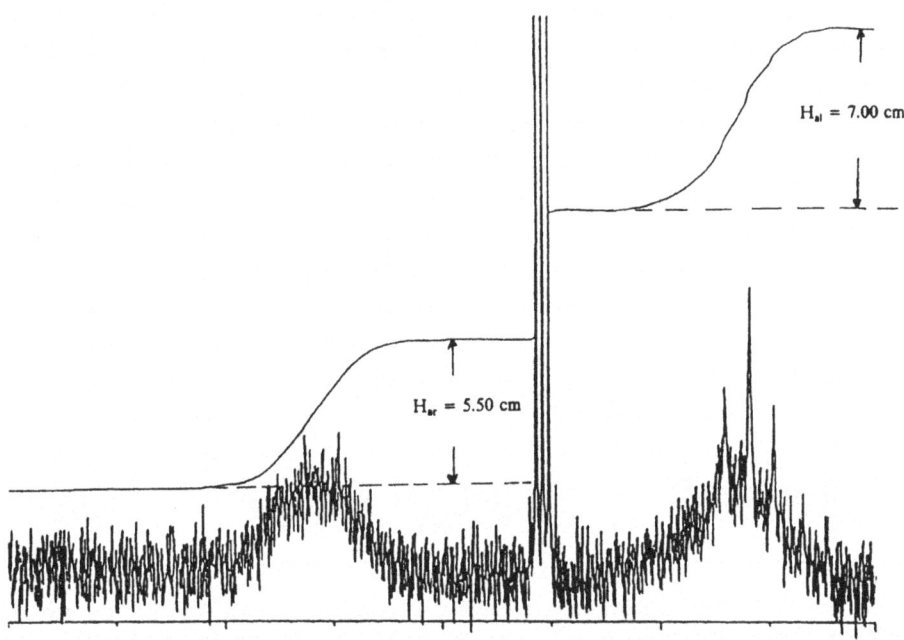

Figure 3. The NMR spectrum of AAA-1 asphaltenes 10.9% by weight in chloroform solution (H_{ar}: the height of aromatics, H_{al}: the height of aliphatics)

Table III. Source and refinery process of the samples

Sample	Source	Refinery Process
AAA-1	Lloydminister	distillation
AAB-1	Wyoming Sour	distillation
AAD-1	California Coast	distillation
AAG-1	California Valley	distillation
AAK-1	Boscan	distillation
AAM-1	West Texas	solvent

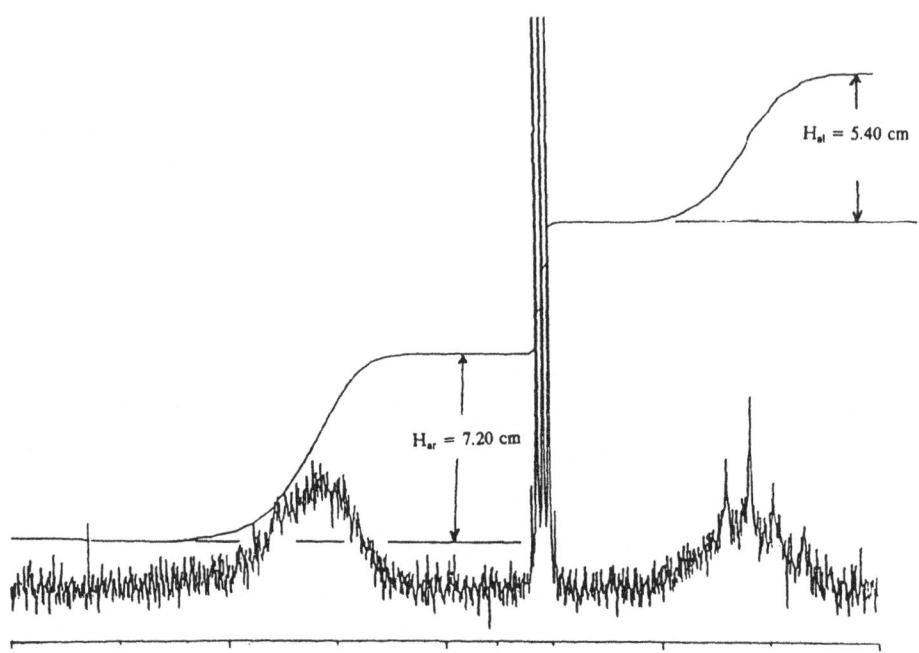

Figure 4. The NMR spectrum of AAB-1 asphaltenes 10.6% by weight in chloroform solution (H_{ar}: the height of aromatics, H_{al}: the height of aliphatics)

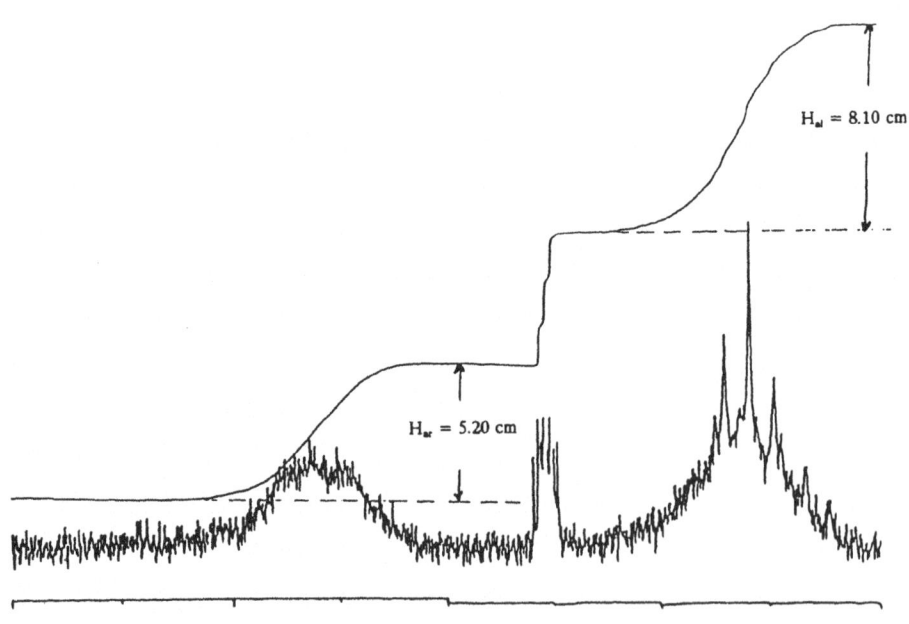

Figure 5. The NMR spectrum of AAD-1 asphaltenes 15.1% by weight in chloroform solution (H_{ar}: the height of aromatics, H_{al}: the height of aliphatics)

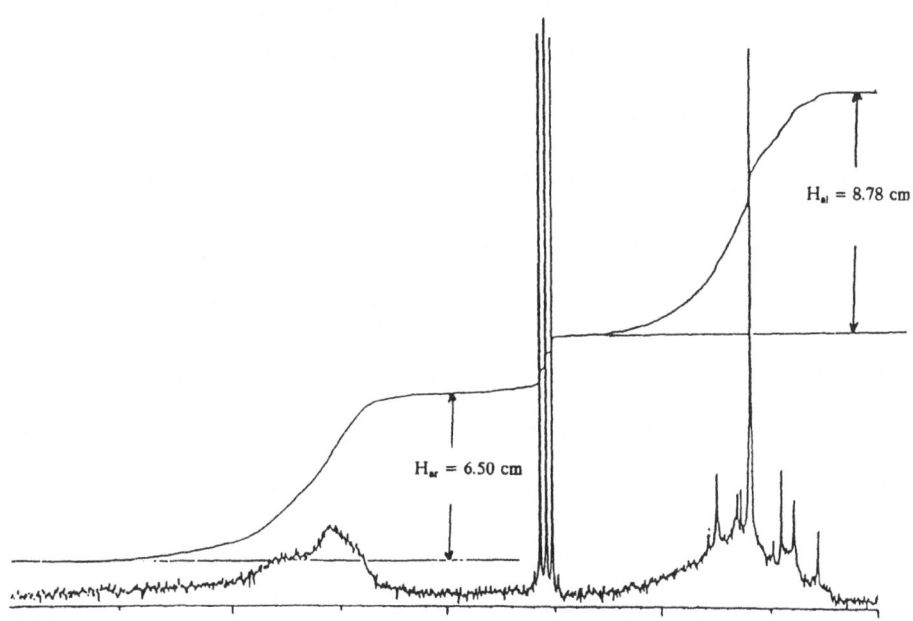

Figure 6. THe NMR spectrum of AAG-1 asphaltenes 15.0% by weight in chloroform solution (H_{ar}: the height of aromatics, H_{al}: the height of aliphatics)

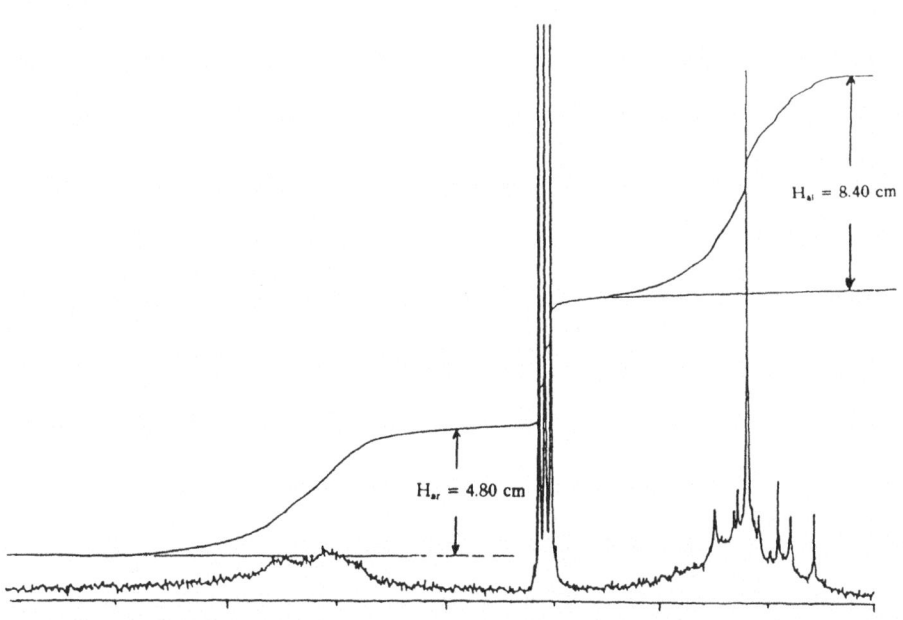

Figure 7. The NMR spectrum of AAK-1 asphaltenes 15.0% by weight in chloroform solution (H_{ar}: the height of aromatics, H_{al}: the height of aliphatics)

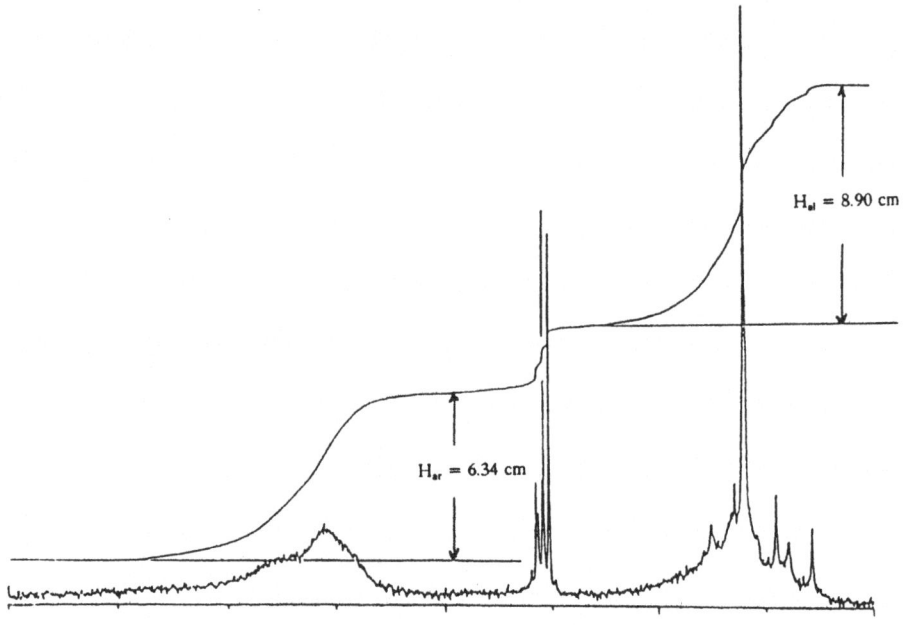

Figure 8. The NMR spectrum of AAM-1 asphaltenes 10.0% by weight in chloroform solution (H_{ar}: the height of aromatics, H_{al}: the height of aliphatics)

Yen's previous data including native petroleum asphalt and non-petroleum asphalt with refinery asphalt was listed in Table V[12]. By plotting at data on Figure 10 the correlation between asphaltene aromaticity and the H/C ratio of asphaltene was obtained. It seems like the asphalts which have slopes are higher than -1 are sol types, while those whose slopes between -1 and -1.5 are sol-gel types, and gel types slopes are lower than -1.5. These results coincide with the refinery asphalt's results in Figure 9. We believe that colloidal types of asphalt can be discerned by asphaltene aromaticity and the hydrogen carbon ratio approach.

Another approach is trying to prove that the native asphaltene aromaticity will increase with the increasing temperature. Figures 11 to 13 show the NMR spectra of AAG-1 asphaltene at different temperatures and each sample's aromaticity can be easily determined by Equation 2. Table VI lists the aromaticities[13] of some native petroleum asphaltene and the aromaticities of AAG-1 asphaltene (from experiment) at different temperatures. Figure 14 indicates that the aromaticities from native petroleum asphaltene increase with temperature, but those from refinery asphalt (AAG-1) do not because the production temperature of refinery asphalt is already above 450°C. Again, we prove that the rheology of asphalt could be changed due to temperature differences especially at high temperatures where aromaticity of asphaltene is increased.

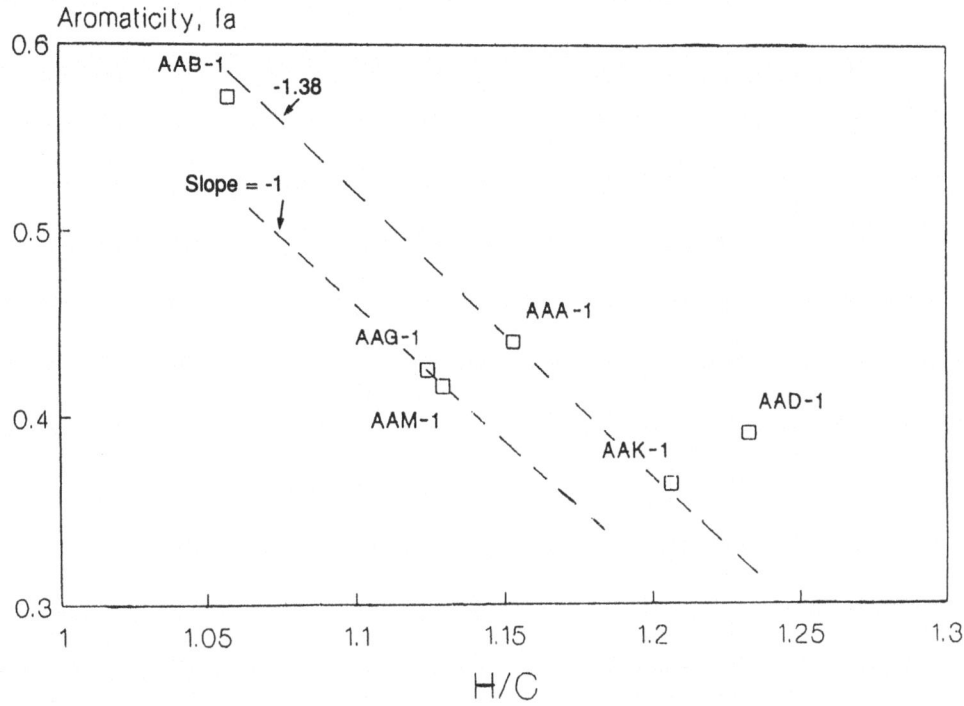

Figure 9. The plot of NMR aromaticity vs H/C ratio for different asphaltenes from six core asphalts

Table IV. Aromaticity vs. H/C atomic ratio of six asphaltenes by NMR

Asphaltene Sample	Har. Aromatics (cm)	Hal. Aliphatics (cm)	f_a Aromaticity Har / (Har + Hal)	H / C
AAA-1	5.50	7.00	0.44	1.15
AAB-1	7.20	5.40	0.57	1.06
AAD-1	5.20	8.10	0.39	1.23
AAG-1	6.50	8.78	0.43	1.12
AAK-1	4.80	8.40	0.36	1.21
AAM-1	6.34	8.90	0.42	1.13

Table V. The asphaltene H/C ratio and aromaticity from different sources

NO.	ASPHALTENE	H/C	fa
Refinery Asphalt			
1	AAA-1	1.15	0.44
2	AAB-1	1.06	0.57
3	AAD-1	1.23	0.39
4	AAG-1	1.12	0.43
5	AAK-1	1.21	0.36
6	AAM-1	1.13	0.42
Native Petroleum Asphalt			
7	Baxterville (a)	1.02	0.51
8	Baxterville (b)	1.05	0.53
9	Bachaquero	1.08	0.41
10	Lagunillas	1.13	0.41
11	Boscan (a)	1.15	0.35
12	Boscan (b)	1.16	0.35
13	Burgan (Kuwait)	1.17	0.38
14	Raudhatain	1.17	0.32
15	Wafra No. A-1	1.18	0.37
16	Melones	1.18	0.38
17	Mara	1.19	0.35
18	Wafra No. 17	1.19	0.35
19	Belridge	1.19	0.30
20	Yorba Linda	1.21	0.38
21	Santiago	1.22	0.31
22	Libya	1.25	0.25
23	Ragusa	1.29	0.26
Non-Petroleum Asphalt			
24	Athabasca (a)	1.21	0.31
25	Athabasca (b)	1.21	0.38
26	Rozel Point	1.40	0.18
27	Coorongite	1.49	0.41
28	Tabbyite	0.92	0.15
29	Grahamite (BS)	1.22	0.31
30	Grahamite (BI)	1.01	0.46
31	Mavjak (BS)	1.04	0.47
32	Mavjak (BI)	1.06	0.48
33	Gilsonite (BS)	1.36	0.21
34	Gilsonite (BI)	1.41	0.29

Table VI. The aromaticity of petroleum and AAG-1 asphaltene at different Temperature

Temperature °C	AAG-1 Asphaltene			Petroleum Asphaltene
	Har. Aromatics (cm)	Hal. Aliphatics (cm)	f_a Aromaticity Har / (Har + Hal)	f_a Aromaticity
25	32.20	71.67	0.30	0.35
156	31.30	76.66	0.29	-
230	30.42	75.21	0.29	-
340	145.46	273.98	0.35	-
350	-	-	-	0.59
380	-	-	-	0.58
490	-	-	-	0.66

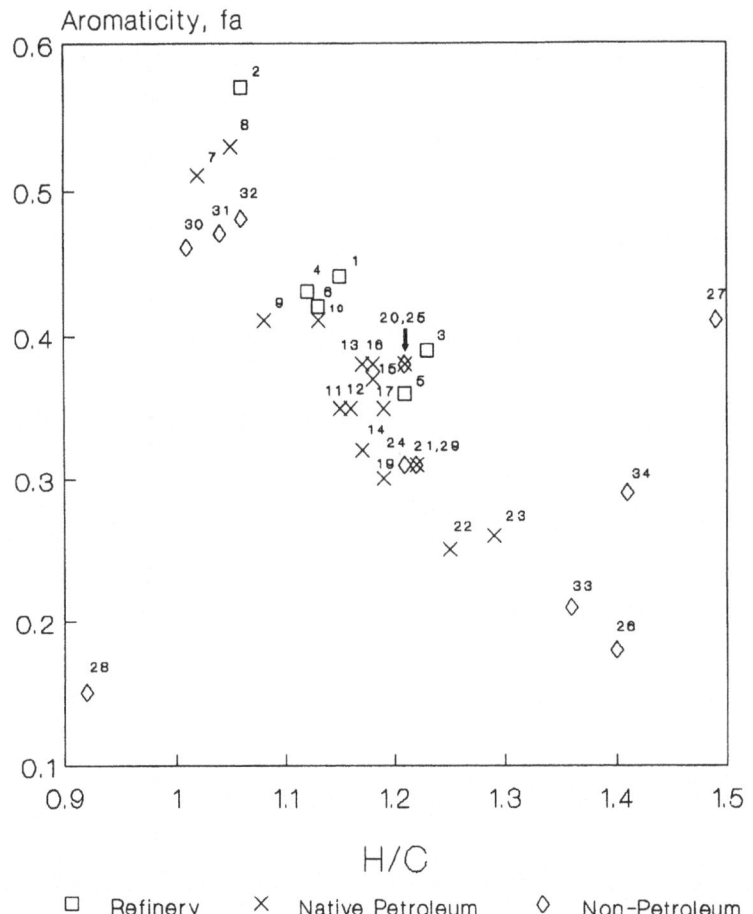

Figure 10. The plot of asphaltene aromaticity vs H/C ratio from various asphalt sources

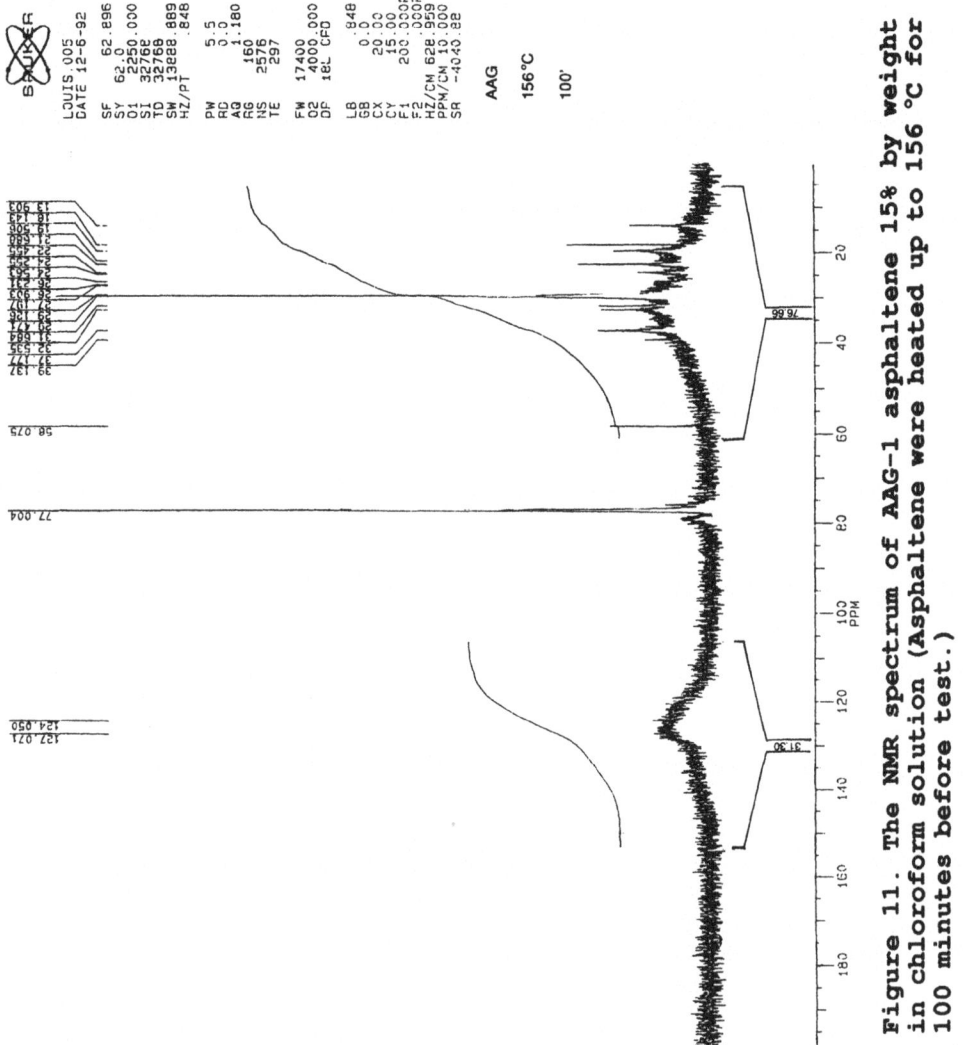

Figure 11. The NMR spectrum of AAG-1 asphaltene 15% by weight in chloroform solution (Asphaltene were heated up to 156 °C for 100 minutes before test.)

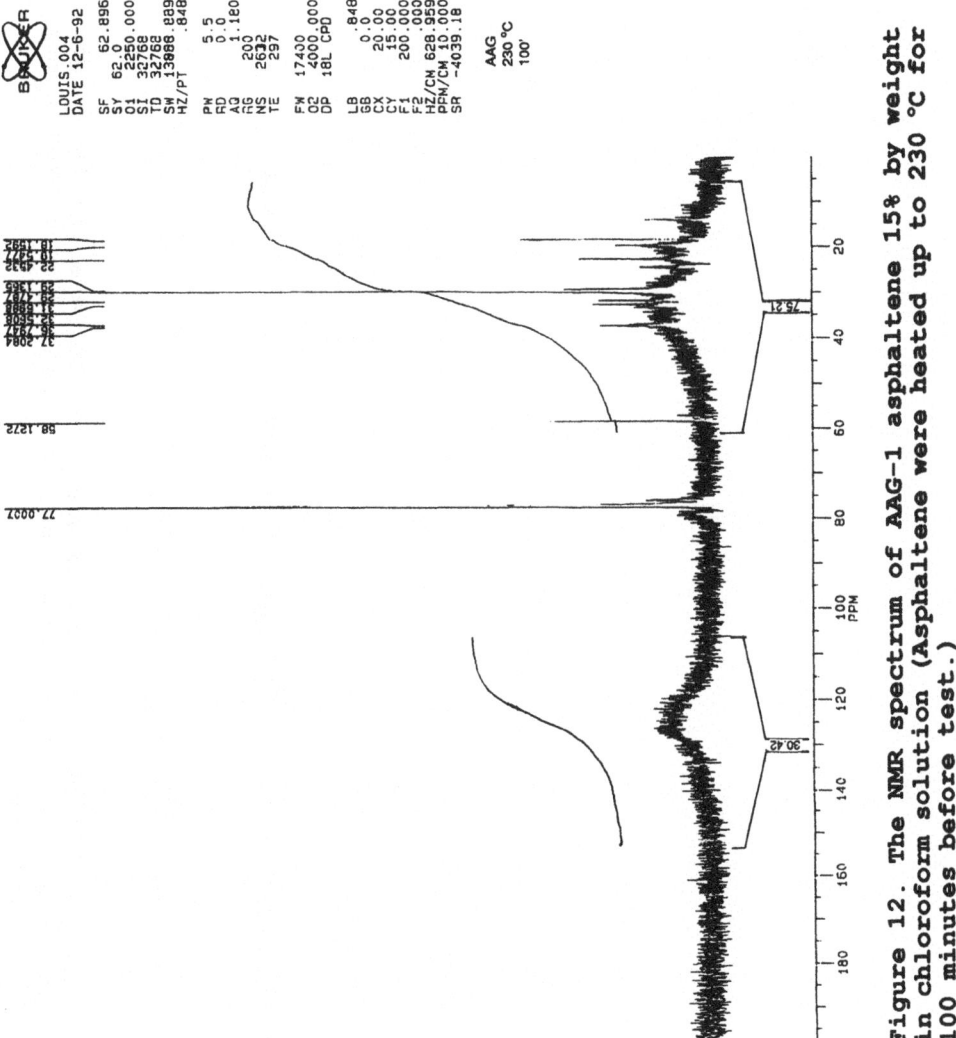

Figure 12. The NMR spectrum of AAG-1 asphaltene 15% by weight
in chloroform solution (Asphaltene were heated up to 230 °C for
100 minutes before test.)

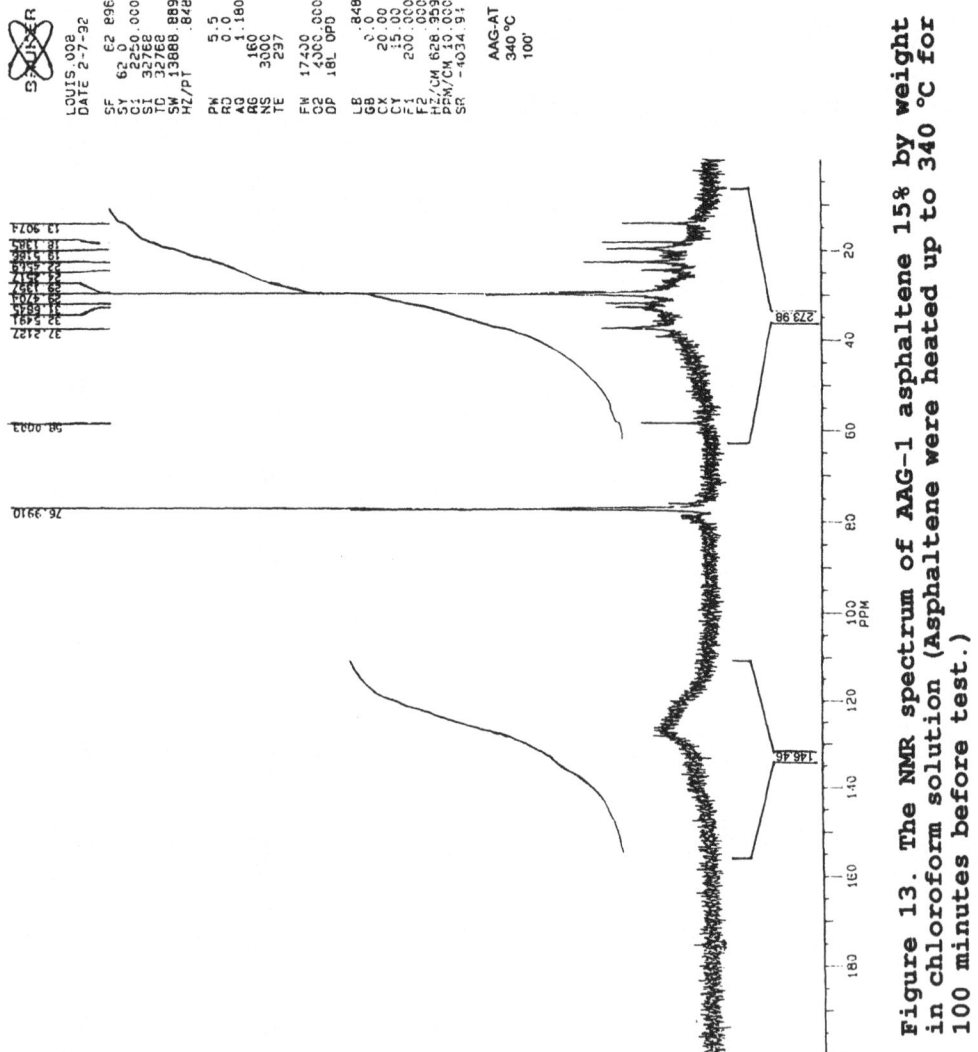

Figure 13. The NMR spectrum of AAG-1 asphaltene 15% by weight in chloroform solution (Asphaltene were heated up to 340 °C for 100 minutes before test.)

CONCLUSION

According to their rheological properties, asphalts can be classified into three different types -sol, sol-gel, and gel types. The penetration index is one of the most popular indexes that distinguishes asphalt into three types because it is easy to calculate by Equation 1 if one has penetration (at 77 °C) and softening point data. The NMR asphaltene aromaticity values can correlate with the H/C ratio of asphaltene. Based on this, the parent asphalt can classify into three types. Temperature change is an important factor affecting asphalt's rheology and characteristics.

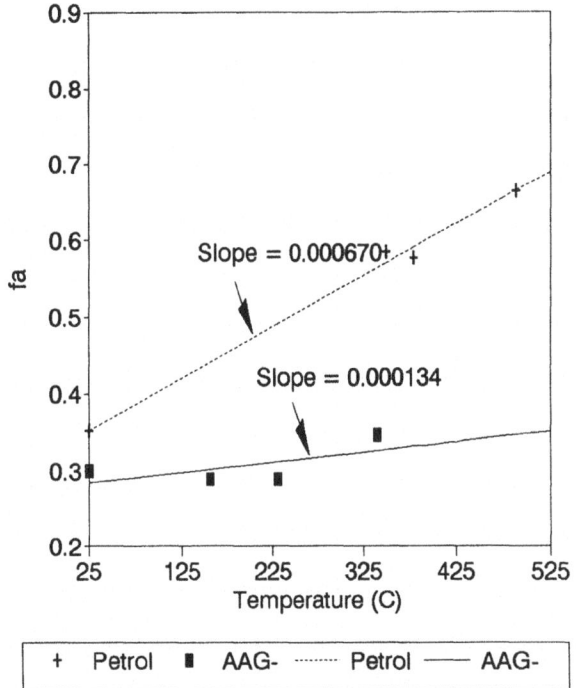

Figure 14. The comparison between native petroleum and AAG-1 asphaltene aromaticity with temperature

ACKNOWLEDGEMENT

The authors want to acknowledge Allan Kershaw of the chemistry department for helping in the NMR work.

REFERENCE

1. ASTM D4124, "Standard Test Method for Separation of Asphalt into Four Fractions", American Society for Testing and Materials, Philadelphia, Pa., (1988).

2. Schwager, J. and Yen, T. F.; Coal Liquefaction Products from Major Demonstration Processes I. Separation and Analysis", Fuel, 57(2), 100-104 (1978).

3. Altgelt, K. H., Jewell,D. M., Latham, D.R. and Selucky, M. L.; "Chromatography in petroleum analysis", In Chromatographic Science Series. ed. K. H. Algelt, Marcel Dekker Inc., New York, NY, pp.194-196 (1979).

4. Lian, H. J. Lee, C., Wang, Y. Y. and Yen, T. F.; Characterization of Asphalt with the Preparative Chromatotron, Journal of Planar Chromatography, 5, 263-266, 1992.

5. Lian, H. J., Lee, C. and Yen, T.F.; Fractionation of Asphalt by Thin-Layer Chromatography interfacial with Flame Ionization Detector (TLC-FID) and Characterization by FTIR, Fine Particle Society Conference, Las Vegas, July (1992).

6. The Asphalt Institute, "Introduction to Asphalt: Manual Series No. 5. Eighth Edition", College Park, MD.

7. Barth, E. J.; "Asphalt:Science and Technology", chap 4, p.252 (1962).

8. Pfeiffer, J. P.; "The Properties of Asphaltic Bitumen", Elsevier Publishing Company, Inc., New York, chap IX, p.239, (1950).

9. Strategic Highway Research Program:Asphalt Contractors Meeting Notes, attachment 4, Denver, Colorado, February 2-3 (1989).

10. Willard, H. H., Merritt, L. L., Jr., Dean, J. A. and Settle, F. A., Jr.; "Instrumental Methods of Analysis (seventh edition)", Wadsworth, Inc., p.422-425 (1988).

11. "Binder Characterization and Evaluation", SHRP A-002A quarterly report, Western Research Institute, pp. 28, March (1989).

12. Yen, T. F., Erdman, J. G. and Pollack, S. S.; Investigation of the Structure of Petroleum Asphaltenes by X-Ray Diffraction, Analytical Chemistry, 33, 1587-1594 (1961).

13. Yen, T. F. unpublished results, (1966).

SLUGE FORMATION DURING HEAVY OIL

CONVERSION

David A. Storm, Stephen J. DeCanio and Eric Y. Sheu

Texaco, Research and Development
P.O. Box 609
Beacon, New York 12508

ABSTRACT

Sludge formation during visbreaking, or the catalytic hydro-conversion of residual sets limits on the amounts of these feed-stocks converted into more useful products. Large amounts of sludge form at high conversions in these refinery processes, and it fouls downstream units. It is generally believed that sludge forms because the asphaltenes flocculate during processing. The exact mechanism is not known, however. The asphaltenes may become less soluble, or they may become less well solubilized, or both during processing. Although it is known that the amount of sludge formed depends on the feed-stock, or crude source, it has not been possible to link feed-stock properties to the tendency to form sludge during processing in a reliable manner. In this work we develop "single molecule" representations of the asphaltenes (heptane insolubles) and resins (pentane insolubles/heptane solubles) from several crude sources based on the elemental composition, and parameters determined by C13 NMR and proton NMR. The amount of sludge formed in laboratory tests can be correlated to three feed-stock parameters: the amount of polynuclear aromaticity in the asphaltenes, the degree of alkyl-subsitution on the polynuclear core in the single molecule representation of the asphaltenes, and the number of "asphaltene molecules" to "resin molecules".

Asphaltene Particles in Fossil Fuel Exploration, Recovery, Refining, and Production Processes, Edited by M.K. Sharma and T.F. Yen, Plenum Press, New York, 1994

81

INTRODUCTION

An important operation in refining is the upgrading of the bottom of the barrel into more valuable products. Unfortunately sludge often forms during these upgrading processes, and fouls downstream equipment, ultimately limiting the conversion or the severity of the process. The amount of sludge appears to depend on the temperature of the processing, and properties of the crude oil used to make the vacuum residue.

It is generally believed that sludge forms because the asphaltenes flocculate during processing. The exact mechanism is not known however. The asphaltenes may become less soluble because of molecular structural changes, or they may become less well solubilized by the surrounding medium because of molecular changes in the medium, or both.

In this work we develop "single molecule" representations of the asphaltenes(heptane insolubles) and "resins"(pentane insolubles/heptane solubles) fractions from several crude oils based on the elemental composition, parameters determined by liquid state C^{13} NMR and proton NMR, and molecular weight information obtained by mass spectroscopy. We then show the amount of sludge formed in laboratory hydro-conversion experiments can be related to three molecular parameters obtained from the "single molecule" representations: the amount of polynuclear aromaticity, the degree of substitution of the polynuclear aromatics by aliphatic groups, and the number of "asphaltene molecules" to "resin molecules". These parameters appear to account for both molecular changes to the asphaltenes, and changes in the solubility relationships between the asphaltenes and the medium during processing in some approximate manner.

EXPERIMENTAL

The vacuum residues used in this work were obtained from the crude oils shown in Table I by vacuum distillation; they have an apparent boiling point greater than 1050 °F. These vacuum residues were hydro-processed in a 300 cc Robinson-Mahoney CSTR in the presence of a presulfided Mo/Al_2O_3 catalyst at 805 °F. The hydrogen pressure was 2250 psig, and conversion to products with boiling points less than 1050 °F was in the range of 60-70%. Samples of the liquid product were collected, and filtered hot to determine the amount of sludge formed during processing.

Asphaltenes, or heptane insolubles, were obtained from the vacuum residues by mixing one part of the vacuum residue with forty parts of heptane, stirring overnight at room temperature, and then filtering the solids. The resins (pentane insolubles/heptane solubles) were obtained from the remaining oil using pentane as the precipitating agent.

TABLE I. Properties of Vacuum Residues

Crude Oil	Location	Wt.% Asph.	Wt % Resins	Sludge-g/h
Duri	Indonesia	3.5	10.6	zero
ANS	Alaska	6.7	8.1	2.84
Ratawi	M.E.	21.8	7.6	1.35
Merey	Venezuela	31.0	6.7	5.13
Oriente	Ecuador	29.0	5.9	5.9

Proton (300 MHz) and carbon-13 (75 MHz) NMR spectra were measured at room temperature using a 5 mm switchable 1H/13C probe on a Varian VXR-300 spectrometer. Approximately 100 mg of the asphaltenes or resins were dissolved in 1 ml of a 0.03 M solution of chromium acetonyl acetonate in deuterated chloroform. Proton spectra were collected using an 11 us pulse width and a 2 s recycle delay. Carbon-13 spectra were collected using a 15.5 us pulse width and a 6 s recycle delay. Gated decoupling was employed during the acquisition to negate NOE effects.

RESULTS AND DISCUSSION

The amounts of asphaltenes and resins obtained from the vacuum residues are listed in Table I along with the amount of sludge obtained in the liquid product after processing. Typical Carbon-13 and proton NMR spectra are shown in Figures 1a and 1b, respectively. Based on previous studies[1,2] five chemically distinct carbon domains were delineated: aliphatic carbon, C(AL), from 10-60 ppm; aromatic carbon, C(AR), from 110-160 ppm; aromatic carbon that is shared by three aromatic rings and aromatic carbon with a hydrogen attached, $C(AR;b_3,H)$, from 110-130 ppm; aromatic carbon shared by two aromatic rings, naphthenic carbon, and aromatic carbon with an attached methyl group, $C(AR;CH_3,b_2,n)$, from 130-138 ppm; and finally aromatic carbon attached to an alkyl-group, C(AR;R), from 138-160 ppm. Similarly four regions were integrated in the proton spectra: aromatic hydrogen, H(AR), from 6-9 ppm; hydrogen which is alpha to an aromatic ring, H(ALPHA), from 2-4 ppm; hydrogen which is beta to an aromatic ring, H(BETA), from 1-2 ppm; and hydrogen which is gamma to an aromatic ring, H(GAMMA), from 0.5-1 ppm.

a

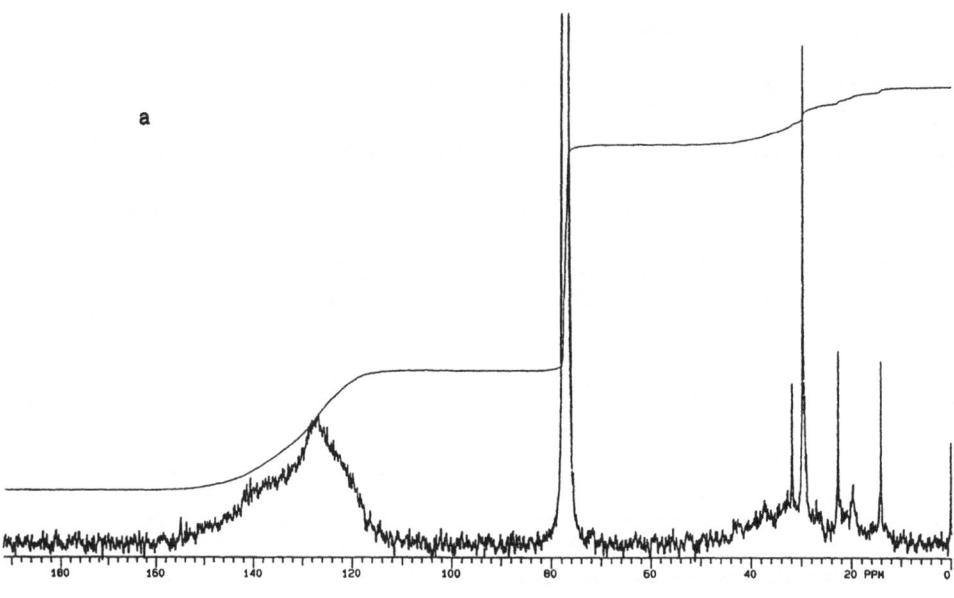

b

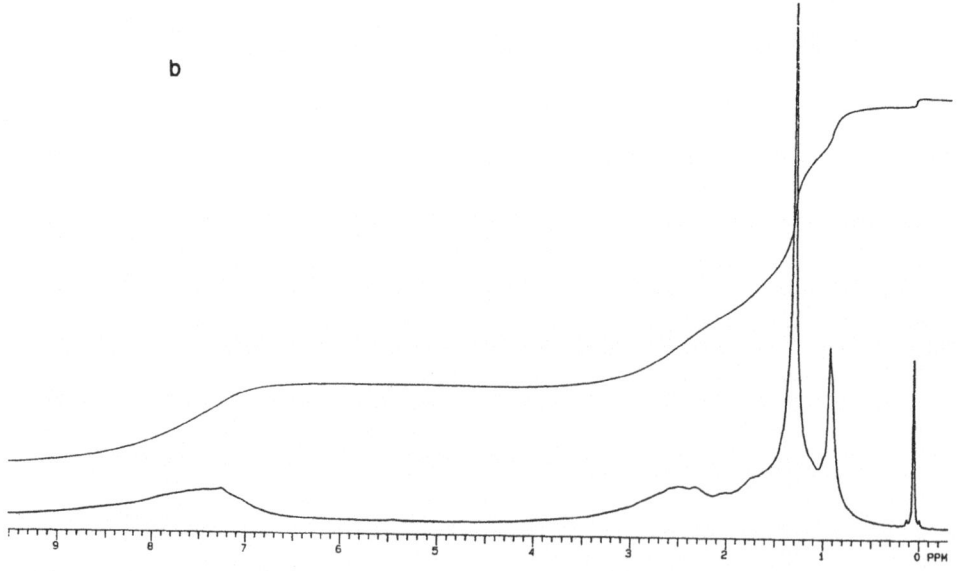

Figure 1. Typical C-13(1a) and Proton(1b) NMR Spectra For Asphaltenes.

The amount of triple bridge carbon, b_3, was calculated by normalizing the H(AR) region to the carbon spectrum and subtracting this area form $C(AR;b_3,H)$ as follows:

$$b_3 = C(AR;b3,H) - H/C \times H(AR) \qquad (1)$$

where H/C is the atomic ratio of hydrogen and carbon determined by standard analytical methods. NMR parameters for the oil, resin and asphaltene sub-fractions of the various vacuum residues are given in Table II.

TABLE II. NMR Parameters for Ratawi Vacuum Residue

Parameter	Oil	Resins	Asphaltenes
H/C	1.6	1.53	1.22
C(AR)	27.8	41.3	58.1
$C(AR;b_3)$	0.0	0.4	9.8
C(AR;H)	10.2	16.2	16.5
C(AR;R)	9.9	11.7	14.1
$C(AR;n,b_2,CH_3)$	7.8	31.5	17.9
C(AL)	72.2	58.7	41.9
C(AL;a)	4.9	3.2	2.3
C(AL;b)	4.0	3.8	2.8
C(AL;d+e)	24.3	19.7	13.1
H(AR)	7.3	10.6	13.7
H(ALPHA)	14.0	12.5	22.0
H(BETA)	59.6	53.3	50.8
H(GAMMA)	19.0	23.6	13.6

It is surprising that one can deduce an organic molecule that matches these NMR parameters fairly easily. The significance of this molecule to asphaltenes will be discussed in the next section. We will illustrate the procedure for Ratawi asphaltenes, although the method worked for all the asphaltenes studied in this work. The procedure is as follows:

(1) Pick a trial polynuclear aromatic(PNA) by finding a PNA that has the same amount of b_3 carbon per aromatic carbon as that measured by NMR. From Table II:

$$b_3/C(AR) = 9.8/58.1 = 0.17 \qquad (2)$$

The PNA shown in Figure 2a appears to be suitable; this PNA contains 19 carbon atoms, 11 of which are on the outside available for bonding to other atoms. The ratio of b_3 carbon to aromatic carbon is 0.16, while that of b_2-carbon/$C(AR)$ is 0.26. The fraction of carbons on the outside($o/C(AR)$)is 0.58.

(2) One then checks that the trial PNA is satisfactory by calculating $b_2/C(AR)$ and $o/C(AR)$ using the NMR parameters. From Table II:

$$b_2/C(AR) = 17.9/58.1 = 0.31 \qquad (3)$$

$$o/C(AR) = (16.5 + 14.1)/58.1 = 0.53 \qquad (4)$$

assuming the amount of $C(AR;CH_3)$ and $C(AR;n)$ is small.

One sees that the trial PNA is satisfactory. The agreement could be made exact if we assumed there was appropriate amounts of $C(AR;CH_3)$ and $C(AR;n)$, but this is not necessary to do at this point.

(3) Calculate the fraction of outside positions substituted by an alkyl-group(R) and the number of R-groups attachments per PNA. From Table II, again assuming that $C(AR;CH_3)$ is small:

$$f = 14.1/(14.1+16.5) = 0.46 \qquad (5)$$

Since there is 11 outside positions per PNA, this implies there are about 5 R-groups attachments per PNA.

(4) Calculate the number of aliphatic carbons per PNA:

$$aliphatics/PNA = C(AL)/C(AR) \; x \; 19 \; =41.9/58.1 \; x \; 19 \; =13.7 \quad (6)$$

Since there are 5 R-groups attachments per PNA, this implies there is 2-3 aliphatic carbons per attachments.

(5) Calculate the number of $-(CH_2)-$ per attachment:

$$-(CH_2)- = C(AL;d+e)/C(AL) = 13.1/41.9 = 0.31 \qquad (7)$$

Since there are 2-3 aliphatic carbons per attachments, this implies there is one $-CH_2-$ group per R-group.

Next we make use of the fact that the number average molecular weight of Ratawi asphaltenes is approximately 800 amu[3]. If one did not have information about the molecular weight, one could simply add five C_3 units to the PNA shown in Figure 2a and obtain a representation consistent with the NMR parameters. The molecular weight would be 444, however. As the number average molecular weight is about 800, we double the structure and deduce the diagram shown in Figure 2b. By construction this diagram represents the NMR parameters, the elemental H/C ratio, and the upper bound on the number average molecular weight[3].

DISCUSSION

Since vacuum residue is a mixture of thousands of molecules[4] it would appear hopeless to try relate detailed molecular properties to processing characteristics. Experience suggests that the some processing characteristics, such as sludge or coke formation, are related to the asphaltenes, and perhaps how they are solubilized by the "resins". "Resins" in this work are not the standard resin fraction, which also includes polar molecules in our "oil" fraction. Unfortunately the asphaltene and resin fractions are also mixtures of thousands of molecules, however[3-5]. Therefore if we are to relate molecular parameters to processing characteristic, we must represent the asphaltenes, and other important sub-fractions, by the molecular information available for the mixtures. This information can be conveniently displayed in diagrammatic form, as in Figure 2b. These molecular diagrams are not molecules as normally used in chemistry. They are simply diagrams that represent a certain amount of analytical data about these fractions. The diagram shown in Figure 2b represents the Ratawi asphaltenes with respect to NMR parameters, molecular weight, and atomic H and C concentrations. It is not a faithful representation, however, because it does not contain any heteroatoms, which are known to be in the asphaltenes.

Not withstanding these drawbacks, we have tried to relate a processing characteristic, the amount of sludge formed in laboratory hydro-processing experiments, to molecular parameters determined from representations for the asphaltenes and resins. The representation of the resin is very simple in this analysis; the resins are represented by the amount present in the vacuum residue, and their upper bound on their number average molecular weight. The representation of the asphaltenes is also simpler that the diagram shown in Figure 2b; they are represented by the

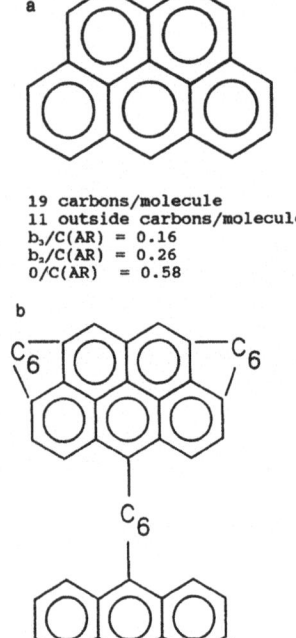

19 carbons/molecule
11 outside carbons/molecule
$b_3/C(AR)$ = 0.16
$b_2/C(AR)$ = 0.26
$0/C(AR)$ = 0.58

Figure 2. Trial PNA For Ratawi Asphaltenes(2a) and Molecular Representation For Ratawi Aspahltenes(2b).

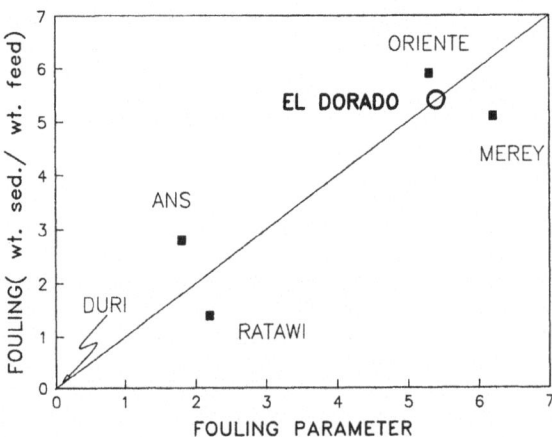

Figure 3. Amount of Fouling Versus The Fouling Parameter.

amount present in the vacuum residue, the upper bound on the number average molecular weight, the fraction of outside positions substituted by R-groups, and the number of b_3-carbons per PNA. Our reasoning is as follows: The amount of sludge formed should be proportional to the number of asphaltene molecules present, which is the amount of material divided by the upper bound on the number average molecular weight. It should also be proportional to the number of resin molecules present, if one assumes the resins solubilize the asphaltenes. Furthermore one might expect that it would be proportional to the size of the asphaltene PNA cluster, given by the b_3 parameter, and how soluble the asphaltene is in the vacuum residue, given by the fraction of outside positions substituted by a R-group. There is some speculation that R-groups on the PNA are related to solubility of the asphaltenes.

Therefore we have fit the following function to the amounts of sludge formed in the laboratory experiments:

$$F = A_1 \ x \ b_3/C(AR) \ + \ A_2 \ x \ f \ + \ A_3 \ x \ RESINS/ASPHALTENES \qquad (8)$$

The fit is shown in Figure 3. The data point marked El Dorado is an independent observation.

CONCLUSIONS

1. One can find a single "trial" polynuclear aromatic that fits the observed liquid state carbon and proton NMR spectra for vacuum residues from several crude oils. One first calculates the fraction of triple bridgehead carbon using the combined carbon and proton spectra, and then finds a hypothetical PNA that has that fraction of triple bridgehead carbon.

2. Such a "single molecule representation" represents the NMR spectra, and it can represent the "average molecule", if that concept has meaning and if the NMR measurement is equally sensitive to all carbon and proton environments in the sample. The "average molecule" may not have a physical meaning, however, because molecular properties change from molecule to molecule in discrete units while the average of a discrete set of properties may not be a possible discrete property(the average of integers is not necessarily an integer).

3. The "single molecule representations" obtained in this procedure satisfy the upper bound shown to hold for the number average molecular weight; i. e. it is not necessary to make an independent assumption about the number average molecular weight to obtain "molecular representations".

4. One can use the molecular parameters obtained from the single molecule representations to correlate sludge formation in residue upgrading processes with feed-stock molecular properties.

REFERENCES

1. Yen, T. F., Wu, W. H., Chilingar, G. V., "A Study of Structure of Petroleum Asphaltenes and Related Substances by Proton Nuclear Magnetic Resonance" Energy Sources, **7(3)**, 275 (1984).

2. Gillet, S., Rubini, P., Delpeuch, J-J., Escalies, J-C., and Valentin, P., "Quantitative Carbon-13 and Proton Nuclear Magnetic Resonance Spectroscopy of Crude Oil and Petroleum Products-I: Some Rules for Obtaining a Set of Reliable Structural Parameters" Fuel, **60**, 221 (1981).

3. Storm, D. A., DeCanio, S. J. and DeTar, M. M., "Upper Bound on Number Average Molecular Weight of Asphaltenes" Fuel, **69**, 735 (1990).

4. Boduszynski, M. M., "Composition of Heavy Petroleum-I: Molecular Weight, Hydrogen Deficiency, and Heteroatom Concentration as a Function of Atmospheric Equivalent Boiling Point Up to 1400^0F" Energy & Fuels, **1**, 2 (1987).

5. DeCanio, S. J., Nero, V. P., DeTar, M. M. and Storm, D. A., "Determination of the Molecular Weights of Ratawi Vacuum Residue Fractions: A Comparison of Mass Spectrometric and Vapour Phase Osmometry Techniques" Fuel, **69**, 1233 (1990).

UPGRADING FROM PETROLEUM AND COAL-DERIVED ASPHALTENES

Jiunn-Ren Lin*, Joon-Kyu Park and Teh Fu Yen

Civil and Environmental Engineering
University of Southern California
Los Angeles, CA 90089 U.S.A.

ABSTRACT

Asphaltene is the major obstacle for the catalytic cracking process in the industrial refining practices as well as coal liquefaction. The generation of coke and the adsorption of asphaltene particles reduce the effective surface area of catalyst. In addition, the efficiency of coal hydropyrolysis is decreased due to the agglomeration of asphaltenes. Thus, most of the conventional hydrogenation processes are working at high temperature and high hydrogen pressure. Ultrasound can provide localized cavitation centers with extremely high temperature and high pressure at room temperature and ambient pressure. Surfactant is widely used in enhance oil recovery and is able to improve the dispersion of asphaltene particles. A solid phase of sodium borohydride is used to provide a constant hydrogen supply in this investigation. With the existence of surfactant and ultrasound, both petroleum asphaltene and coal-derived asphaltene can be converted into lower molecular compounds (gas oil and resins) with a satisfactory conversion. The performance of the commercially available surfactant Span 20 is compared with the self-generated surfactant Wet-sol on both petroleum and coal-derived asphaltenes.

*Present Address: Energy and Resource Labs, Industrial Technology Research Institute, Chutung 310, Hsinchu, Taiwan

Asphaltene Particles in Fossil Fuel Exploration, Recovery, Refining, and Production Processes, Edited by M.K. Sharma and T.F. Yen, Plenum Press, New York, 1994

INTRODUCTION

Asphaltenes, either petroleum or coal asphaltene, consist of high-molecular-weight polyaromatic or polycyclic nuclei with linkage of heteroatoms such as N, O, and S.[1,2] Due to its polar nature, asphaltene tends to agglomerate into larger aggregates. The existence of asphaltenes is the major obstacle for oil upgrading as well as coal liquefaction.

Catalytic cracking process has typically been the most effective conventional method for converting heavier petroleum oils into lighter and more valuable products. However, asphaltenes tend to deactivate the catalyst through the creation of coke on fixed bed catalyst and through excessive coke on reformation equipments. Thus, these processes have historically limited success in processing residue or heavy oils because the high proportion of asphaltenes and concomitant levels of contaminants are usually found in the oils. In order to overcome the Laplace pressure for the oil layers and enhance the reaction rate, oil upgrading process is generally operated at high temperature and/or high pressure.

The fluid fuels are desirable during coal hydropyrolysis. Thus, hydrogen and partially hydrogenated coal-derived liquids are participated during the process. The addition of hydrogen is to increase the H/C ratio and the yield of coal tar. The hydrogenated coal-derived liquid serves as hydrogen source and transfer agent for degrading coal. Coal asphaltene is recognized as the mediate materials of coal liquefaction process.[3] Both asphaltenes and coal particles have the tendency of aggregation, and reduction of the efficiency of hydrogenation process. To overcome the attraction forces and facilitate the hydrogenation, similar to oil upgrading, the coal liquefaction process is performed at a high temperature and high pressure. The high temperature and high pressure create the constraints and limitations on the reactor vessel materials for safety consideration. These drawbacks can be solved when ultrasonic process is applied to perform these processes at ambient temperature and atmospheric pressure.

The reactions initiated by ultrasonic irradiation began 50 years ago. Ultrasound energy may activate various mechanisms to promote the effects, but the mechanism involved are not always known or understood. So far, most investigators agree that there are three phenomena attribute the effects caused by ultrasonic irradiation. First, a rapid movement of fluids caused by a variation of sonic pressure causes solvent compression and rarefaction. Second and by far the most important, is cavitation. Most investigators accept that the formation and collapse of microbubbles are responsible for most of the significant chemical effects observed. Third, there is microstreaming in which a large amount of vibrational energy is confined into small volumes with very ittle heating.

Experimentally, cavitation thresholds up to 300 bars (3.0×10^7 Pa) have been obtained.[4] The cavitation temperatures measured are as high as 5200±650 K and 5075±156 K for alkane solvents and silicone oil, respectively.[5,6] The concentration of microbubbles in aqueous solution of 5% human serum albumin would be as high as 3 billion microbubbles per mL when irradiated with ultrasound under air for 3 minutes.[7] In addition, the mean diameter of microbubbles is around 5 μm. The violent implosion of the microbubbles also leads to luminescence. The high temperature and pressure result in the molecular collision. For a 20 KHz ultrasonic unit, the period per cycle is 5.0×10^{-5} second. The collision of water molecules happens in less than one quarter of the cycle. Thus, the collision time should have an order of magnitude of about 1.25×10^{-5} second.

Under these cavitation conditions, thermal scission of the bonds of the heavy fraction occurs. The asphaltene sheet will break into lighter fractions (gas oils and/or resins). In addition, through the effect of cavitation, the association and the aggregation of asphaltene are reduced. In such a manner that the oil upgrading and coal liquefaction can be enhanced.

Surfactants are amphiphatic molecules that include both distinct hydrophobic and hydrophilic regions. These molecules can be formed and isolated from an asphaltene micelle.[8,9] Heavy fraction molecules, including resins and asphaltenes, will form either Hartley micelles or reversed micelles from their interaction with surfactants.[2,10] During the rearrangement of the molecules in these micelles, the functional groups are likely to orient toward the aqueous phase. This particular orientation facilitates the reaction with radicals from cavitation centers. Thus, selecting a proper surfactant will ensure the stability of asphaltene dispersion.

The ultrasound has been used with surfactant to upgrade and/or recovery of the fossil fuels at ambient temperature and atmospheric pressure. The preliminary studies of tar sand, asphalt, heavy oil, coal liquid and oil shales show this process is feasible.[11] During these studies, we found that Span 20 is the best commercial available surfactant in the selected surfactants. The Wetsol is the spent solution which surfactant is self-generated from a mixture of sodium silicate, hydrogen peroxide, and tar sand during the sonication process. Both of these two surfactants are tested and compared in this investigation to convert both petroleum and coal asphaltene to lighter fractions. The results show that the Wetsol has better performance than that of 0.01% Span 20 on both Boscan asphaltene and Catalytical coal asphaltene but worse on Synthoil asphaltene.

EXPERIMENTAL

There are one petroleum asphaltene and two coal-derived asphaltenes used in this investigation. Petroleum asphaltene was isolated from the Boscan asphalt which was obtained through the material bank of the Strategic Highway Research Program (SHRP). One of the coal-derived liquid was obtained from Pittsburgh Energy Research Center, Department of Energy #IS-72-5 by Synthoil process, the other is from Catalytic Inc., Wilsonville #IS-71-3 by Solvent Refined Coal (SRC) process. The isolation of asphaltene and two other coal asphaltenes are respectively from Boscan Petroleum an two coal liquid samples. The method is done by the Soxhlet process by n-pentane as the eluent for more than 48 hours. The asphaltene after Soxhlet process was dried overnight in the vacuum oven at room temperature.

Span 20 and Wet-sol are the surfactants used in this study to improve the dispersion of asphaltenes. Span 20 is a nonionic surfactant which is commercially available. The detailed procedure has been reported previously and is briefly described as follows:[8] Wetsol is the spent solution of a mixture of sodium silicate, hydrogen peroxide and tar sand after the sonication process. Tar sand sample used is a high grade Athabasca tar sand, from McMurray Formation. A typical example is that 200g of tar sand are applied to 1000 ml of alkaline solution and sonicated for 30 minutes with the addition of few drops of 30% hydrogen peroxide as free-radical initiators. The alkaline solution is made by mixing 20:1 by volume distilled water to sodium silicate and 200 mL of 0.02 N sodium hydroxide. Stirring was applied to help mixing the solution. In this experiment, sonication was performed in a 10-gallon 40 KHz transducerized tank. The transducers deliver approximately 2.0 W/cm^2 acoustic intensity at the transducer surface. After reaction, the bitumen was skimmed from the surface of solution. Since the surfactant in the spent the solution will form micelle and even vesicle, the fresh spent solution is used to perform the further experiments within two hours after sonication process.[15]

For each run, 0.50g of asphaltene was dissolved in the 1.0g of toluene first, 0.01% of Span 20 and 30g of distilled water (or 30g of Wetsol solution) were added gradually in a 50ml flask. After adding materials into 50ml flask, to increase the interfacial area, vigorous stirring was applied during the water or Wetsol addition utilized to form an oil-in water emulsion. A model VC-50 ultrasonic processor (Sonics & Materials, Inc.) at output power level of 8 W, 20MHz was used in the process. 0.4g of sodium borohydride was added every 15 minutes reaction time.

After reaction, asphalt constituent was extracted by the dichloromethane. Both 30ml of dichloromethane and the sample after reaction are put into the separation funnel and shaken several times. Two layers appear in a shortly. Since part of the emulsion is quite stable and hard to extract from the oil phase, shaking several times is required at that point. The dichloromethane with asphalt formed the bottom layer. After 15 minutes, the bottom layer in the separation funnel was collected for further analysis. The material balance showed that the average recovery from the extraction process was above 98%. The soluble portion of organic matter in the aqueous phase usually was less than 2%. Since analysis of the asphalt content is dependent on the four fraction ratios, part of the product remaining in the water phase will not significantly affect the analysis result. The dichloromethane with asphalt will stay in the bottom layer.

The composition of asphalt and coal samples were determined through the TLC/FID analysis. Before starting the analysis, a blank scan of Chromarod is required to ensure the removal of contaminants. Around 1 μL of sample was spotted on the Chromarod-SIII made by Iatron Laboratories, Inc. Since humidity would significantly affect the results and reproductivity of TLC development, preserving the Chromarod in the 25% constant humidity chamber using sulfuric acid for at least 10 minutes to reach equilibrium is required. Two solvents with different polarities, including toluene and a mixture of dichloromethane and methanol at a ratio of 95 to 5, were used to develop the sample into three fractions (two fractions for coal samples). The oil fraction was expanded first in the TLC development chamber by using toluene as an eluent. Using the mixture of dichloromethane and methanol as an eluent will expand the resin fraction. Finally, the asphaltene fraction remained the same, due to its polarity.

After development, the Chromarods were mounted on the rack and the analyzed by the Itatroscan TH-10 analyst. Hydrogen pressure and air flow rate used for FID analysis were 0.9 in Hg and 2,000 cfm, respectively. The HP 3390A integrator was connected to obtain the composition for three different fractions. Data are collected within 1 minute for each sample. Since a small amount of asphaltene fraction will move into resin and oil during the TLC developing process, internal standardization and calibration were necessary.

RESULTS AND DISCUSSION

The reactions of hydrocarbons initiated by ultrasound have been demonstrated and described as the Rice mechanism.[5,12,13] According to the Rice mechanism, the cracking of asphaltenes could be classified as free radical reactions. Generally, three categories of free radical reactions can be obtained for oil upgrading by ultrasound: initiation, propagation and termination. Asphaltenes could form free radicals with lower molecular weight by bond cleavage under ultrasound. The free radical reactions could be terminated by recombination or disproportionation.

From the reaction mechanisms, several elements are required to convert asphaltene into gas oil and resins.[14] Energy is indispensable for overcoming the chemical bonds to form smaller molecules. Ultrasound could achieve this goal by providing localized high temperature and pressure through the cavitation in the water and organic solution system.

Hydrogen radical supply is another important element for oil upgrading. The hydrogen radicals could terminate the free radical reactions after bond cleavage as well as saturate the products. Since the generation of hydrogen radicals from the dissociation of water are limited, the hydrogen supply is necessary. In this investigation, a solid reducing agent, sodium borohydride, is introduced as a hydrogen source. Sodium borohydride reacts with water and constantly provides hydrogen gas.

Surfactants have widely been applied to reduce the interfacial tension and change the physical properties of system. It may enhance the transportation of free radicals between the two phases. Thus, the free radicals will be easier to transfer from the water phase to the organic phase and stay in micelle. In addition, the surfactant will be adsorbed on the asphaltene particles and prevent the agglomeration of particles.

Figures 1 and 2 represent the conversion of oils and resins from Boscan asphaltenes at reaction time up to 60 minutes. The results show that both oil and resin contents increase with reaction time. They are consistent with the data reported previously.[14] In addition, those studies revealed that the conversion of oil and resin depended not only on the species of surfactant but also on the content of surfactants. Span 20 has been reported to be the best selected surfactant for oil upgrading from petroleum asphaltene; Wetsol is a self-generated surfactant which is efficient for tar sand recovery and upgrading. Figures 1 and 2 show that around 20% and 25% of Boscan asphaltene has been converted into oils and resins within 60 minutes when 0.01% Span 20 or Wetsol are participated, respectively. It appears that the performance of Wetsol has a better performance than 0.01% Span 20 in this system. However, the surfactant content in the WetSol is unable to be determined, the performance of these two surfactant is incomparable and not much conclusion can be stated. In addition, from the trend of asphaltene decrease, it appears that the system with WetSol solution has not reached equilibrium after reacting for 60 minutes. The asphaltene content still has a tendency to decrease. On the other hand, the reaction within the system with surfactant Span 20 almost occurs within the first 20 minutes which is consistent with the previous reports with various asphalt samples. It may imply that the surfactant will affect the mechanism of reactions.

Figure 3 and Table I represent the results of two coal-derived asphaltenes with various surfactants after sonication for 15 and 60 minutes, respectively. Within 15 minutes, around 20 percent of coal-derived asphaltenes has been converted into oils and resins for both samples when Wetsol is participated. Again, it demonstrated the ability of process by ultrasound with surfactant on the oil upgrading. Comparing the yield of oil and resin to petroleum and coal-derived asphaltenes, indicates that the Wetsol solution has a consistent performance. After 15 and 60 minutes, the yield of oil and resin is around 20 and 25 percent for all three samples. However, the yields are vary significantly when Span 20 is in the system. It may imply that the surface characteristics of coal-derived asphaltene and petroleum asphaltene are different. From the results for Synthoil and Catalytical asphaltene, it shows that

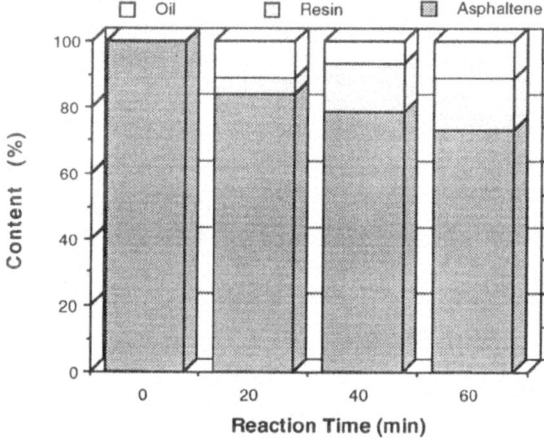

Fig. 1. The conversion of Boscan asphaltene to oil and resin at various reaction time with the participation of Wet-sol.

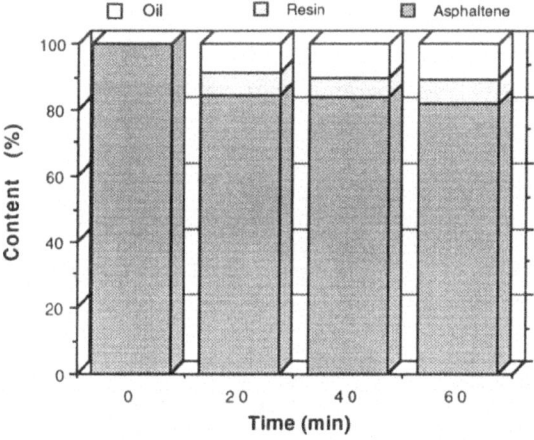

Fig. 2. The conversion of Boscan asphaltene to oil and resin at various reaction time with the participation of 0.01% of Span 20.

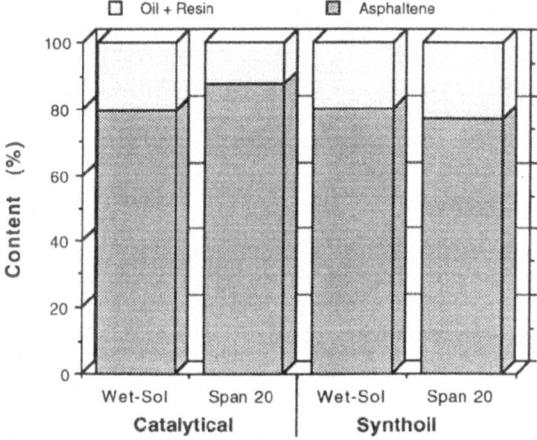

Fig. 3. The composition of two coal-derived asphaltene with various surfactant after sonication for 15 minutes.

Wetsol has better performance for former sample than 0.01% Span 20, but it is worse on the latter sample. It may imply that the characteristics of asphaltene from various source and process are different.

Table I. The composition of coal-derived asphaltene with various surfactant after sonication for 60 minutes.

| | Content (%) | | | |
| | Catalytical | | Synthoil | |
	Wet-sol	Span 20	Wet-sol	Span 20
Oil + Resin	25.6	17.2	25.3	31.1
Asphaltene	74.4	82.8	74.7	68.9

CONCLUSIONS

It has been demonstrated that the process of ultrasound with surfactant can be used to convert both petroleum and coal-derived asphaltenes into gas oil and resin. The process can overcome the drawbacks of conventional processes to be performed at ambient temperature and atmospheric pressure.

The results for Boscan asphaltene reveal that surfactant may affect the mechanism of reactions. The system with WetSol solution indicates continuous decrease on asphaltene content even after 60 minutes. Whereas, the system with surfactant Span 20 almost occurs within first 20 minutes.

For both Boscan and Catalytical asphaltenes, it appears that the performance of Wetsol is better than that of 0.01% Span 20. However, the Synthoil asphaltene has opposite trend. It may imply that the characteristics of asphaltene from various source and process are different. In addition, the yields are various significantly when Span 20 is in the system. It may imply that the surface characteristics of coal-derived asphaltene and petroleum asphaltene are different.

Even though the performance of Span 20 varies for three asphaltene samples, Wetsol solution has a consistent performance. Since Wet-sol is a self-generated surfactant from tar sand, it may consist of several surfactants with different structures.

ACKNOWLEDGEMENTS

We would like to thank Texaco Oil Company for the partial support.

REFERENCES

1. Yen, T. F.; Structural Differences between Asphaltenes Isolated from Petroleum and from Coal Liquid. in: "Chemistry of Asphaltene" (edited by Bunger, J.W. and Li, N.C.), American Chemical Society, Washington D.C., pp39-51 (1981).

2. Yen, T. F.; Asphaltic Materials, in; "Encyclopedia of Polymer Science and Engineering," Second edition, John Wiely & Sons, New York, pp1-10,(1990).

3. Shalabi, M. A., Baldwin, R. M., Bain, R. L., Gary, J. H. and Golden, J. O.; Noncatalytic Coal Liquefaction in a Donor Solvent. Rate of Formation of Oil, Asphaltenes, and Preasphaltenes, Industrial and Engineering Chemistry, Process Design and Development, **18(3)**, 474-479 (1979).

4. Briggs, L. J.; Limiting Negative Pressure of Water, Journal of Applied Physics, **21(7)**,721-722 (1950).

5. Suslick, K. S., Hammerton, D. A., and Cline, R. E.; The Sonochemical Hot Spot, Journal of the American Chemical Society, **108**, 5641-5642 (1986).

6. Flint, E. B. and Suslick, K. S.; The Temperature of Cavitation, Science, **253**, 1397-1399 (1991).

7. Grinstaff, M. W. and Suslick, K. S.; Air-Filled Proteinaceous Microbublles: Synthesis of an Echo-Contrast Agent, Precedings of the National Academic Science of the United States of America **88(17)**, 7708-7710 (1991).

8. Sadeghi, K. M., Sadeghi, M. A. and Yen, T. F.; Novel Extraction of Tar Sands by Sonication with the Aid of In-Situ Surfactants, Energy and Fuels, **4(5)**, 604-608 (1990).

9. Sadeghi, K. H., Sadeghi, M. A., Kuo, J. F., Jang, L. K., Lin, J. R. and Yen, T. F.; A New Process for Tar Sand Recovery, Chemical Engineering Communications (in press).

10. Lian, H. J., Lin, J. R. and Yen, T. F.; Peptization Studies of Asphaltene in Asphalt Systems and Correction by Solubility Parameter Spectra, in: "Particle Technology and Surface Phenomena in Minerals and Petroleum"(edited by sharma, M.K. and Sharma, G.D.) 23-30, Plenum Press, New York (1992).

11. Sadeghi, K. M., Lin, J. R. and Yen, T. F.; Sonochemical

Treatment of Fuel Components, Preprints, American Chemical Society, Division of Fuel Chemistry, **37**, 86-91 (1992).

12. Suslick, K. S., Gawienoski, J. J., Schubert, P. F. and Wang, H. H.; Alkane Sonochemistry. Journal of Physical Chemistry, **87(13)**, 2299-2301 (1983).

13. Suslick, K. S. (editor); "Ultrasound: Its Chemical, Physical and Biological Effect," VCH Publishers, Inc., New York (1988).

14. Lin, J .R. and Yen, T. F.; An Upgrading Process Through Cavitation and Surfactant, Energy and Fuels (in press)

15. Yen, T. F., Park, J. K., Lee, K. I. and Li, Y., Fate of Surfactant Vesicles Surviving from Thermophilic Halotolerant, Spore Forming, Clostridium Thermohydrosulfuricum, _in_: "Microbial Enhancement Oil Recovery," (edited by Domaledson, E.C.) Elsevier Science, Amsterdam, pp297-309 (1990).

CHEMICAL ASPECTS OF ENVIRONMENTALLY

PROCESSED ASPHALT

James L. Conca and Stephen M. Testa*

Environmental Sciences
Washington State University Tri-Cities
100 Sprout Road, Richland, Washington 99352

*Applied Environmental Services, Inc.
23113 Plaza Pointe Drive (Suite 100)
Laguna Hills, California 92653

ABSTRACT

Several aspects of asphalt such as road pavement properties, sensitivity to moisture damage, leaching behavior and functional group analysis and asphalt composition have been discussed. These investigations indicate that the asphalted contaminated soil exhibits high stability and adequate performance as an end product. Asphalt with high contents of pyridinic, phenolic and ketone groups exhibits best chemical stability and performance. It is recommended that the use of monovalent salts and high ionic strength solutions in asphalt cements is avoided because the decrease in chemical stability of the asphalt cement due to disruption of the functional group-aggregate bonds resulting increase in overall permeability.

INTRODUCTION

The use of asphalt cement to stabilize contaminants in petroleum hydrocarbon-affected contaminated soil (PHAS) using Environmentally Processed Asphalt (EPA) remedial technology has been shown to be viable and creative method of utilizing PHAS to produce an useful end-product instead of a waste requiring disposal[1-3]. It is now necessary to demonstrate that contaminants, especially metals, will be retained by the asphalted PHAS. The testing and performance assessment of asphalts has traditionally focused on structural performance as pavement and building materials. The presence of petroleum-contaminated soil as part of the aggregate will not affect the structural behavior of the asphalt as long as general requirements are met, e.g. less than 10% of the aggregate in total volume can be fines. Consequently, evaluating the long-term performance as asphalt liners produced with petroleum-contaminated soil as part of the aggregate in a disposal situation must focus on the chemical performance and requires more extensive experimental study. The petroleum products and asphalt components are generally immiscible with respect to the aqueous phases expected under subsurface conditions with the exceptions of polar functional groups, helping to prevent the release of contaminants from the asphalt and slowing diffusion of aqueous components out of the environmentally processed asphalt. Therefore, the release of metals and other components from the environmentally processed asphalt is coupled with the hydraulic conductivity, diffusivity and structural breakdown of the asphalt itself, which are all slow processes.

For the same crude source and chemical treatments, the end product asphalt cement formed from a cold-mix emulsion or from a hot-mix has the same general structure, composition and properties such as leachability[3]. The greatest effects between the different preparations will be on kinetically-controlled reactions and sorption processes. One might expect that contaminant transport in an asphalted PHAS could be enhanced during the evaporative phase of the cold-mix process as some of the void structure collapses and water leaves the system. However, leaching studies indicate that this is not the case[3].

ASPHALT-AGGREGATE CHEMISTRY

Asphalt is a material with a complex and poorly understood chemistry and structure that depend upon the crude petroleum source and any chemical treatment and/or chemical modifiers added during processing[4]. Asphalt has a large number of hetero-atomic groups with a wide range of chemical reactivities[5]. Therefore, it seems likely that a number of asphalt-metal-soil reactions that could affect contaminant mobility might occur in PHAS subjected to Environmentally Processed Asphalt methodology. The functional groups found in asphalts are shown in Figure 1. The number and distribution of these groups vary widely among different asphalts and determine much of their chemical behavior and performance. Carboxylic acids, ketones and anhydrides are generally formed by atmospheric oxidation (ketones and anhydrides) or caustic pre-treatment (carboxylic acids concentrated in treated asphaltic residues), and are rare in fresh asphalts. However, these may be important in waste disposal situations or EPA methodologies because of possible interactions with oxidizing or alkaline contaminant solution, or if recycled asphalt is used.

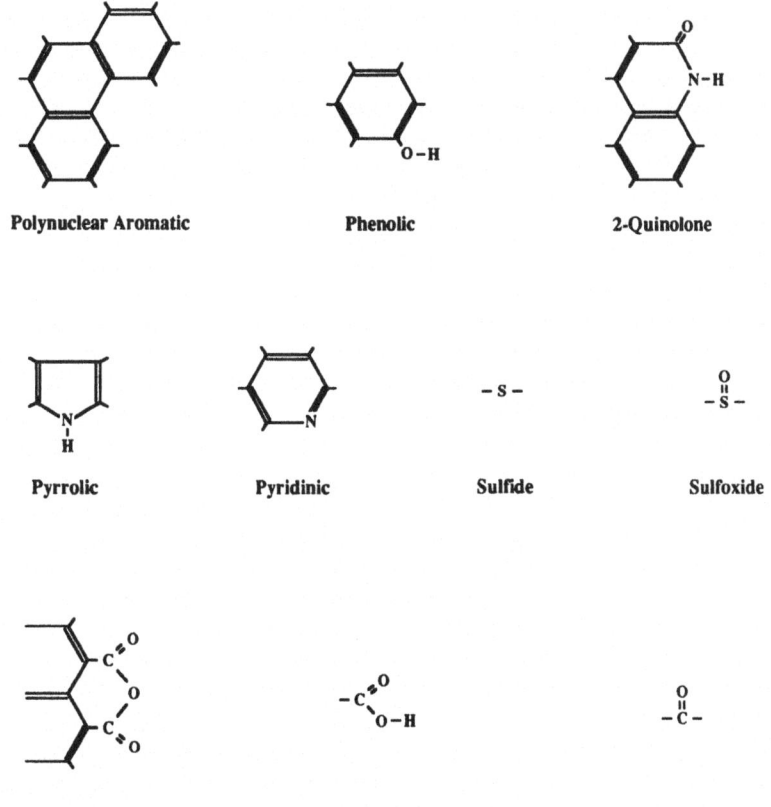

Figure 1. Important functional groups present in asphalt.

There has been no comprehensive study of asphalt
chemistry in relation to the aggregate or to contaminant
species, therefore, generalizations about the chemical
performance of these systems is difficult. However,
extrapolations from asphalt studies of road pavement
properties, leaching behavior, sensitivites to moisture-
damage and functional group analysis[4,6-10] have provided
information that can be used to evaluate the stability of
metals and contaminants in contaminated soils that have been
asphalted.

Generally, it is assumed that asphalt cements are
colloidal systems made up of a suspension of asphaltene
micelles in an oily medium. Micelles are units of various
molecules, usually organic with minor inorganic components,
that have distinct structural and/or chemical properties.
Micelles play an important role in asphalt behavior. Figure
2 illustrates the structural and chemical components in
contaminated soil that has been asphalted. The figure is
approximately to scale and represents the results and
implications of separate studies of asphalt components. The
major components are the aggregate mineral grains, the bulk
aqueous phase, the sorbed electric double layer on the
mineral surface, the resin-peptized asphaltene micelles, the
polar micelles in the asphalt, the asphalt pore spaces, the
asphalt oily medium, and the asphalt functional groups at
the interfaces between these different components. Each
component will be described separately in reference to
Figure 2.

The aggregate mineral grain portrayed in Figure 2 is a
clay mineral. Clay minerals are important because of their
reactive surfaces and their ability to exchange cations from
the interlayer sites with contaminant ions in solution.
There are two major types of complexing functional groups
associated with silicate mineral grains. The most important
is the siloxane ditrigonal cavity, which occurs in
tetrahedral silicate sheets and gives the clay minerals
their exchange capacities[11]. However, the most abundant
surface functional group is the inorganic hydroxyl groups
exposed on tetrahedral silica, called silanols. Those
exposed on octahedral alumina are called aluminols. There
can be more than one type of surface hydroxyl on a given
mineral surface with different reactivities. The siloxane
complexes only positive ions and groups whereas the surface
hydroxyl complexes both anion and cation species depending
on solution composition and pH. Asphalt functional groups as
well as contaminant species will interact with these surface
complexes.

Figure 2 illustrates several mineral surface
interactions; a strong complexation with a uranyl anion
complex, a weak complexation with an hydrated calcium
complex, an exchange of americium with an interlayer cation,
sorption of some short chain hydrocarbons at the mineral
surface, and bonding of the asphalt quinoline, pyrrolic,
phenolic and carboxylic acid functional groups.

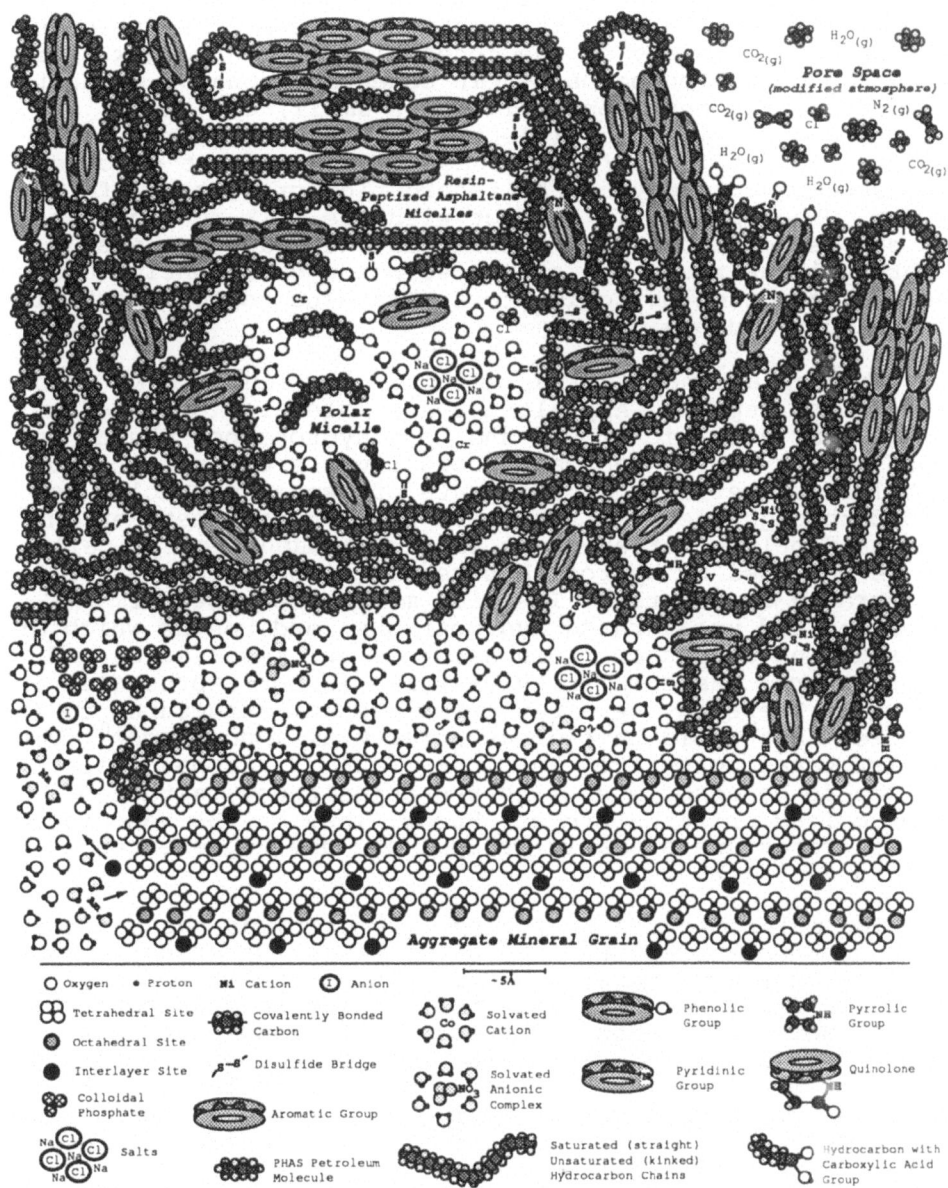

Figure 2. The Asphalt/Aggregate Interface at the Molecular Level Using Space-filling and Structual Representations to Illustrate Important Functional Relationships.

The bulk aqueous phase has many of the dissolved inorganic constituents of interest to contaminant transport. Because the simple diffusion coefficient of dissolved species in the bulk fluid is approximately 10^{-5} cm^2/s regardless of species[12], diffusion of contaminants through the bulk aqueous phase should be the primary route of contaminant release from the asphalt. However, the release rate depends strongly on the connectivity of the bulk fluid, i.e., the diffusion porosity. In coherent asphalts without moisture-damage this connectivity is small, and the effective diffusion coefficient in the asphalt cement[13] as a whole is low (10^{-12} cm^2/s). Release of contaminant species also depends strongly on the retardation properties of the system. Other phases in contact with the bulk aqueous phase can sorb species that are present in the aqueous phase. Figure 2 illustrates the strong sorption of strontium and the weak sorption of iodine to a phosphate colloid particle in the bulk aqueous phase.

The asphaltene micelles and the oily medium make up the bulk of the asphalt. Asphaltenes are molecules composed of polynuclear aromatic groups and long hydrocarbon chains and contain much of the inorganic constituents of asphalt. An asphaltene micelle is an aggregate of asphaltene molecules bonded through pie-electron cloud interactions between the condensed polynulear aromatic sheets[10]. Asphaltene micelles can be peptized by hydrocarbon resins into aggregates as illustrated in Figure 2. The oily medium consists of saturated and unsaturated hydrocarbons, and resins which are a mixture of terpenes, resin alcohols and resin acids and their esters, the complexity of which is not represented in Figure 2. There may be extensive secondary structures to the micelles that could have great structural importance[10]. The metals, vanadium, nickel and iron, found in asphalts are thought to be associated with the asphaltenes, although exact molecular sites are not known. They may be associated with sulfides and aromatic groups, and are represented as such in Figure 2.

The polar micelle in Figure 2 illustrates the possible polar compounds present in polar micelles. Peterson[5] found evidence that agglomerates or micelles of polar asphalt molecules exist separately in the asphalt and sequester certain polar complexes and acids, e.g., manganese-acid complexes. Certainly, carboxylic acid groups will occur in the polar micelles if they exist in a particular asphalt, and polar functional groups such as the phenols, ketones and sulfoxides, will be concentrated at the interface. This sequestering chemically removes those species from subsequent reactions in the non-polar phases and aqueous phases of the asphalt. Speculation exists that water molecules and salts can occur in the polar micelles and this has been incorporated into Figure 2. The polar micelles could sequester many metals species of interest.

The nature of the pore space in asphalt is a highly debated topic. There is a great deal of pore space in asphalt that can be seen in electron micrographs[4] but is not well-connected. Depending upon the conditions of formation and subsequent history, the pore space may or may not be water filled. The gaseous components in pore spaces will include atmospheric components, introduced during formation and processing, and volatile asphalt compounds. Gaseous contaminants of interest may be incorporated into the pore spaces.

Knowing the structures and components of asphalted PHAS is crucial to understanding its chemical performance with respect to metal and petroleum contaminants. As some examples: (1) heavy metal salts of carboxylic acids $(RCOO^-M^+)$, which are insoluble in water, are soluble in strong acids and in strong polar solvents[14], making polar micelles possible repositories of contaminant metal-carboxylic acid salts in the asphalted PHAS; (2) petroleum contaminants in the PHAS will be strongly hydrogen-bonded to the asphalt components as well as possibly bonded through reactions with the functional groups; and (3) oxidation-reduction reactions of metals will be affected by the polar and non-polar phases in the asphalted PHAS. Fish and Elovitz[15] found that the chromium in immiscible cresol/water mixtures preferentially entered the non-aqueous cresol phase (isomeric phenols) where the oxidative-reduction reaction $Cr(VI)$ - $Cr(III)$ occurred much faster than in the aqueous phase. The effect has important implications for metals in PHAS, especially chromium interactions with asphaltic phenolic groups. However, these reactions[15] are kinetically controlled and may be unimportant at pH=8 or so.

ASPHALT-AGGREGATE STABILITY

The chemical stability and environmental performance of the asphalted PHAS depends upon the nature and extent of the asphalt-aggregate bonds. Much of the information concerning these bonds comes from studies of asphalt sensitivity to moisture-induced damage[8]. Because of the polar/non-polar and hydrophobic/hydrophilic interactions among the various phases in the system, the functional groups in Figure 1 will be concentrated at the interfaces between phases, i.e., the asphalt-aggregate interface, the polar micelle-asphalt oil interface, the asphalt-water interface, etc. The behavior of functional groups at the asphalt-aggregate interface determines the chemical stability of the asphalt to a large degree. At the asphalt-aggregate interface, these groups are susceptible to interactions with the aqueous phases and any contaminants present in the PHAS. Petersen[5] determined the relative tendency of the asphalt functional groups to be concentrated at the asphalt-aggregate interface, in order of decreasing tendency, as:

carboxylic acid >> anhydride > quinoline,phenolic > sulfoxide, ketone > pyrrolic

The sorption affinity of the functional groups with the aggregate surface has also been investigated[23] and is given, in order of decreasing tendency, as:

> **pyridinic, carboxylic acid >> anhydride $\geq$ quinoline, phenolic > sulfoxide > ketone >> pyrrolic, polynuclear aromatic**

The third important tendency of the functional groups affecting environmentally processed asphalt stability is the ability of the functional group-aggregate bond to be displaced by water. This tendency, listed in order of decreasing tendency, is given as:

> **anhydride, quinoline, carboxylic acid > sulfoxide > ketone > pyrrolic, phenolic**

Combining these tendencies provides some indications of performance and some guidelines for using environmentally processed asphalt methodology. For the best performance, the asphalt should have high contents of pyridinic, phenolic and ketone groups, which can be achieved by carefully choosing the source material or using additives such as shale oil. Also, the presence of inorganic sulfur, monovalent salts and high ionic strength solutions in the asphalt decreases the chemical stability of the asphalt cement by disruption of the functional group-aggregate bonds and increases the systems's permeability[23]. Addition to lime to the aggregate is often used to counter this effect.

CONTAMINANT MOBILITIES IN ASPHALT

Contaminant mobility, especially of metals, in the asphalted cement will be affected by many factors as follows:

1. diffusivity and permeability of the asphalt cement as a whole (generally less than 10^{-12} cm^2/s and 10^{-9} cm/s, respectively[13])

2. solubility of species in the various aqueous, polar and non-polar phases,

3. speciation of the contaminants, e.g., UO_2 (non-mobile) vs $UO_2.nCO_3$ (highly-mobile),

4. complexation with any chelating organics,

5. redox reactions of metals across aqueous-organic phase boundaries,

6. sorption on aggregate surfaces, along asphalt-aqueous interface, or on colloids,

7. precipitation of solid phases and/or colloids of metal salts, especially oxyhydroxide and carboxylic acid salts.

Many of these properties are not known to the degree that specific contaminant release can be predicted. Leaching tests are the primary method of evaluating contaminant mobility in these systems and have been performed on a variety of asphalts. However, the low diffusivities and permeabilities of asphalt is obviously the greatest factor in the retention of contaminants in asphalt cements. Conditions which adversely affect the diffusivity and permeability will have the greatest adverse effect on contaminant mobility and release. The asphalt acts primarily as a physical containment to the contaminants and the aggregates. Of course, the physical properties of the asphalt cement result from the composition and structure of the asphalt.

CONTAMINANT LEACHING FROM ASPHALT

Even though metals such as vanadium and nickel occur in asphalt at the hundreds of parts per million levels[4] as well as many toxic organic components, asphalt leachates and products have never produced toxic or contaminated solutions that are considered to be hazardous materials by the Environmental Protection Agency. According to Kriech[16], asphalt's non-volatile, visco-elastic properties results in the general observation that asphalt leachates do not contain contaminant concentrations above the Environmental Protection Agency's drinking water guidelines. Even in asphalts made with metal slags as an aggregate, metals do not become solubilized and do not leach from these asphalts in detectable concentrations even when used with strongly acidic and alkaline solutions[17]. Some typical test results for leaching of asphalt are shown in Table I for several asphalts[16,17].

Asphalt leachates that have contained detectable concentrations of contaminants have been obtained in studies of asphalted nuclear wastes[18-21]. However, in all of these studies, high concentrations of salts were used, as high as 50% salt/50% asphalt. It is known that high salt concentrations in asphalt mixes disrupts the asphalt structure and is a condition that will not occur in PHAS subjected to environmentally processed asphalt methodology.

On the other hand, in diffusion studies of radioactive wastes with normal salt contents, diffusion coefficients were measured to be as low as for normal asphalt conditions[9], with diffusion coefficients of 10^{-12} cm^2/s and 10^{-13} to 10^{-10} cm^2/s. In all studies, researchers point out that experimental effects, e.g., slicing of thin asphalt membranes, may introduce errors that are not relevant to the filed situation and tend to increase the observed diffusion coefficient[22,23].

A number of unplanned leaching experiments have been taking place with asphalt. Asphalt has been used for years to line domestic water reservoirs, especially in California, and to line fish-rearing ponds, with no adverse effects. There are over 30 asphalt-lined fish-rearing ponds in Oregon

Table 1. Leach Test Results From Seven Reclaimed Pavement Asphalts Using Environmental Protection Agency Test Procedures SW846-3350, -8080, -1311, -3510, -8310, and -3010 (after Kriech, 1991, Heritage Research Group). The Symbol < Denotes Below the Given Detection Limit. All Values are Well Below Drinking Water Standards and RCRA Guidelines.

Compound	1	2	3	4	5	6	7
Barium	< 0.2 ppm	0.40 ppm	0.36 ppm	0.33 ppm	< 0.2 ppm	< 0.2 ppm	< 2 ppm
Cadmium	< 0.2 ppm	< 0.2 ppm	< 0.2 ppm	< 0.2 ppm	< 0.2 ppm	< 0.2 ppm	<0.02 ppm
Chromium	<0.05 ppm	0.52 ppm	<0.05 ppm	<0.05 ppm	<0.05 ppm	<0.05 ppm	0.10 ppm
Lead	< 0.2 ppm	1.80 ppm	< 0.2 ppm	< 0.2 ppm	< 0.2 ppm	< 0.2 ppm	< 0.2 ppm
Silver	<0.04 ppm	<0.04 ppm	<0.04 ppm	<0.04 ppm	<0.04 ppm	<0.04 ppm	<0.04 ppm
Arsenic	<0.005 ppm	<0.005 ppm	<0.005 ppm	<0.005 ppm	<0.005 ppm	<0.005 ppm	<0.005 ppm
Selenium	<0.025 ppm	<0.025 ppm	<0.025 ppm	<0.025 ppm	<0.025 ppm	<0.025 ppm	<0.005 ppm
Mercury	<0.005 ppm	<0.005 ppm	<0.005 ppm	<0.005 ppm	<0.005 ppm	<0.005 ppm	<0.005 ppm
1,4 Dichlorobenzene	< 50 ppb	< 50 ppb	< 50 ppb	< 50 ppb	< 50 ppb	< 50 ppb	< 12 ppb
2,4 Dinitrotoluene	< 50 ppb	< 50 ppb	< 50 ppb	< 50 ppb	< 50 ppb	< 50 ppb	< 12 ppb
Hexachlorobenzene	< 50 ppb	< 50 ppb	< 50 ppb	< 50 ppb	< 50 ppb	< 50 ppb	< 12 ppb
Hexachlorobutadiene	< 50 ppb	< 50 ppb	< 50 ppb	< 50 ppb	< 50 ppb	< 50 ppb	< 12 ppb
Hexachloroethane	< 50 ppb	< 50 ppb	< 50 ppb	< 50 ppb	< 50 ppb	< 50 ppb	< 12 ppb
Nitrobenzene	<250 ppb	<250 ppb	<250 ppb	<250 ppb	<250 ppb	<250 ppb	< 12 ppb
Pyridine	<120 ppb	<120 ppb	<120 ppb	<120 ppb	<120 ppb	<120 ppb	< 60 ppb
Cresylic Acid	< 50 ppb	< 50 ppb	< 50 ppb	< 50 ppb	< 50 ppb	< 50 ppb	< 30 ppb
2-Methyl Phenol	< 50 ppb	< 50 ppb	< 50 ppb	< 50 ppb	< 50 ppb	< 50 ppb	< 30 ppb
3-Methyl Phenol	< 50 ppb	< 50 ppb	< 50 ppb	< 50 ppb	< 50 ppb	< 50 ppb	< 30 ppb
4-Methyl Phenol	<250 ppb	<250 ppb	<250 ppb	<250 ppb	<250 ppb	<250 ppb	< 30 ppb
Pentachlorophenol	<250 ppb	<250 ppb	<250 ppb	<250 ppb	<250 ppb	<250 ppb	< 60 ppb
2,4,5-Trichlorophenol	< 50 ppb	< 50 ppb	< 50 ppb	< 50 ppb	< 50 ppb	< 50 ppb	< 30 ppb
2,4,6-Trichlorophenol	< 50 ppb	< 50 ppb	< 50 ppb	< 50 ppb	< 50 ppb	< 50 ppb	< 30 ppb
Naphthalene	0.49 ppb	<0.13 ppb	<0.13 ppb	0.30 ppb	<0.13 ppb	<0.13 ppb	0.25 ppb
Acenaphthylene	<0.20 ppb	<0.20 ppb	<0.20 ppb	<0.20 ppb	<0.20 ppb	<0.20 ppb	<0.15ppb
Acenaphthene	0.14 ppb	<0.13 ppb	<0.13 ppb	<0.13 ppb	<0.13 ppb	<0.13 ppb	<0.194 ppb
Fluorene	<0.015 ppb	<0.015 ppb	<0.015 ppb	<0.015 ppb	<0.015 ppb	<0.015 ppb	<0.023 ppb
Phenanthrene	<0.13 ppb	<0.13 ppb	<0.13 ppb	<0.13 ppb	<0.13 ppb	<0.13 ppb	<0.033 ppb
Anthracene	<0.017 ppb	<0.017 ppb	<0.017 ppb	<0.017 ppb	<0.017 ppb	<0.017 ppb	<0.015 ppb
Fluoranthene	<0.017 ppb	<0.017 ppb	<0.017 ppb	<0.017 ppb	<0.017 ppb	<0.017 ppb	<0.037 ppb
Pyrene	<0.060 ppb	<0.060 ppb	<0.060 ppb	<0.060 ppb	<0.060 ppb	<0.060 ppb	<0.04 ppb
Benzo(A)Anthracene	<0.017 ppb	<0.017 ppb	<0.017 ppb	<0.017 ppb	0.017 ppb	<0.017 ppb	<0.048 ppb
Chrysene	<0.033 ppb	<0.033 ppb	<0.033 ppb	<0.033 ppb	<0.033 ppb	<0.033 ppb	<0.017 ppb
Benzo(B)Fluoranthene	<0.023 ppb	<0.023 ppb	<0.023 ppb	<0.023 ppb	<0.023 ppb	<0.023 ppb	<0.02 ppb
Benzo(K)Fluoranthene	<0.017 ppb	<0.017 ppb	<0.017 ppb	<0.017 ppb	0.050 ppb	<0.017 ppb	<0.022 ppb
Benzo(A)Pyrene	<0.240 ppb	<0.240 ppb	<0.240 ppb	<0.240 ppb	<0.240 ppb	<0.240 ppb	<0.023 ppb
Dibenzo(A,H)Anthracene	<0.068 ppb	<0.068 ppb	<0.068 ppb	<0.068 ppb	<0.068 ppb	<0.068 ppb	<0.018 ppb
Benzo(G,H,I)Perylene	<0.110 ppb	<0.110 ppb	<0.110 ppb	<0.110 ppb	<0.110 ppb	<0.110 ppb	<0.036 ppb
Indeno(1,2,3-CD)Pyrene	<0.022 ppb	<0.022 ppb	<0.022 ppb	<0.022 ppb	<0.022 ppb	<0.022 ppb	<0.021 ppb

and Washington. Trace metal and organic contamination is highly toxic to fry and developing fish. Yet, no adverse effects have been observed from the asphalt liners, indicating a high degree of chemical stability with respect to aqueous solutions, and an absence of any toxicity effects.

CONCLUSIONS

Asphalt studies of road pavement properties, leaching behavior, sensitivities to moisture-damage and functional group analysis have provided information that can be used to evaluate the stability of metals and contaminants in contaminated soils that have been asphalted. These studies indicate that asphalted contaminated soil will be highly stable and will perform adequately as an end product. For the best chemical performance, the asphalt should have high contents of pyridinic, phenolic and ketone groups, which can be achieved by carefully choosing the source material. If the situation requires special stability or redundancy, small amounts of shale oil and lime can be used as additives. Situations and conditions which favor the presence of inorganic sulfur, monovalent salts and high ionic strength solutions in the asphalt should be avoided because these conditions decrease the chemical stability of the asphalt cement by disruption of the functional group-aggregate bonds and by increasing the overall permeability. However, these conditions are not expected in the anticipated uses of asphalt cement to stabilize contaminants in petroleum hydrocarbon-affected contaminated soil using environmentally processed asphalt remedial technology.

ACKNOWLEDGEMENTS

The authors would like to thank Judith Wright for steadfast scientific and editorial review, and Jeff Serne and Henry Plancher for insightful discussions.

REFERENCES

1. Testa, S. M., Patton, D. L. and Conca, J. L.; The Use of Environmentally Processed Asphalt as a Contaminated Soils Remediation Method, in Hydrocarbon Contaminated Soils and Groundwater, Kostecki and Calabrese, Eds., Lewis Publisher (IN PRESS).

2. Preston, R. L. and Testa, S. M.; Permanent and Development Conference on the Soils; Proceedings of the National Research and Development Conference on the Control of Hazardous Materials, Hazardous Materials Control Research Institute, pp.4-10, 1991.

3. Testa, S. M.; Environmental Aspects Relating to Redevelopment of Oil Field Properties within the California Regulatory Framework; Proceedings of the

National Research and Development Conference on the
Control of Hazardous Materials, Hazardous Materials
Control Research Institute, pp.129-132, 1991.

4. Wolfe, D. L., Armentrout, D., Arends, C. B., Baker,
    H. M., Plancher, H. and Petersen, J. C.; Crude Source
    Effects on the Chemical, Morphological and Viscoelastic
    Properties of Styrene/Butadiene Latex Modified Asphalt
    Cements, Transportation Research Record 1096, TRB,
    National Research Council, Washington, D.C., pp.12-21,
    1986.

5. Petersen, J. C.; Quantitative Functional Group Analysis
    of Asphalts Using Differential Infrared Spectrometry and
    Selective Chemical Reactions-Theory and Applications,
    Transportation Research Record 1096, TRB, National
    Research Council, Washington, D.C., pp. 1-11 (1986).

6. Benedetto, A. T., Lottman, R. P., Cratin, P. D. and
    Ensley, E. K.; Asphalt Adhesion and Interfacial
    Phenomena, Highway Research Record No. 340, National
    Research Council Highway Research Board, Washington,
    D.C., 1970.

7. Haxo, J. E., Jr.; Assessing Synthetic and Admixed
    Materials dor Liner Landfills, In Gas and Leachate from
    Landfills: Formation, Collection and Treatment, Eds.
    E. J. Genetelli and J. Circello, Report EPA 600/9-76
    -004, U. S. Enveronmental Protection Agency, Cincinnati,
    Ohio, NTIS Report PB 251161, pp. 130-158, 1976.

8. Petersen, J. C., Plancher, H., Ensley, E. K., Venable,
    R. L. and Miyake, G.; Chemistry of Asphalt-Aggregate
    Interactions: Relationship with Pavement Moisture Damage
    Prediction Test, Transportation Research Record 843,
    TRB, National Research Council, Washington, D.C.,
    pp. 95-104, 1982.

9. Daive, Ch. T. and Vassilev, G. P.; On the Diffusion of
    ^{90}Sr from Radioactive Waste Bituminized by the Mould
    Method, Journal of Nuclear Materials, vol. 127, 132-136,
    1985.

10. Brule, B., Ramond, G. and Such, C.; Relationships
    Between Composition, Structure and Properties of Road
    Asphalts: State of Research at the French Public Works
    Central Laboratory, Transportation Research Record 1096,
    TRB, National Research Council, Washington, D.C.,
    pp. 22-34, 1986.

11. Sposito, G.; The Surface Chemistry of Soils, Oxford
    University Press, New York, 1984.

12. Conca, J. L. and Wright, J. V.; Aqueous Diffusion
    Coefficients in Unsaturated Materials, Scientific Basis
    for Nuclear Waste Management XIV, Materials Research
    Society Symposium Proceedings, vol. 212, pp.879-884,
    1991.

13. Hickle, R. D.; Impermeable Asphalt Concrete Pond Liner, Civil Engineering, pp.56-59, August, 1976.

14. Morrison, R. T. and Boyd, R.; Organic Chemistry, Allyn and Bacon, Inc., Boston, 1974.

15. Fish, W. and Elovitz, M. S.; Cr(VI) Reduction by Phenols in Immiscible Two-Phase Systems: Implications for Subsurface Chromate Transport, Transactions of the American Geophysical Union, 71, 1719, 1990.

16. Kriech, A. J.; Evaluation of RAP for Use as a Clear Fill, Heritage Research Group Report HRG#122EM02, Indianapolis, IN, 1991.

17. Kriech, A. J.; Evaluation of Hot Mix Asphalt for Leachability, Heritage Research Group Report HRG#3959AOM3, Indianapolis, IN, 1990.

18. Amarantos, S. G. and Petropoulos, J. H.; Certain Aspects of Leaching Kinetics of Solidified Radioactive Waste - Laboratory Studies, DEMO 81/2, Greek Atomic Energy Commission, Athens, Greece (1981).

19. Daugherty, D. R., Pietrzak, R. F., Fuhrmann, M. and Columbo, P.; An Experimental Survey of the Factors that Affect Leaching from Low-Level Radioactive Waste Forms, BNL-52125, Brrokhaven National Laboratory, Upton, New York, 1988.

20. Fuhrmann, M., Pietrzak, R. F., Franz, E. M., Heiser, J. H., III. and Columbo, P.; Optimization of the Factors that Accelerate Leaching, BNL-52204, Brookhaven National Laboratory, Upton, New York (1989).

21. Nikiforov, A. S., Zakharova, K. P. and Polyakov, A. S.; Physicochemical Foundations of Bituminization of Liquid Radioactive Wastes from a Nuclear Power Plant with RBMK Reactor and the Properties of the Compound Formed, Soviet Atomic Energy, 61, 664-668, 1987.

22. Eschrich, H.; Properties and Long-Term Behavior of Bitumen and Radioactive Waste-Bitumen Mixtures, SKBF KBS Technical Report No. 80-14, Swedish Nuclear Fuel and Waste Management Company, Stockholm (1980).

23. Partnership, M. R. M.; Bituminous and Asphalic Membranes for Radioactive Waste Repositories on Land, Report to the Department of the Environment, DOE/RW/87.009, Briston, England, 1988.

SURFACE ACTIVITY AND DYNAMICS OF ASPHALTENES

Eric Y. Sheu, M. M. De Tar and D. A. Storm

Research and Development Department
Texaco
P.O. Box 509
Beacon, New York 12508

ABSTRACT

A series of surface tension measurements were performed to study the surface activities as well as the dynamics of Ratawi asphaltenes in organic solvents. Nitrobenzene and pyridine were found to be good candidates for this study. The results indicate that asphaltene molecules are surface active and thermodynamically reversible during the micellization and dissociation processes, similar to surfactant solutions. Due to the structural complexity, the micellization and dissociation kinetics are considerably slower than surfactant systems. This study provides solid evidence that asphaltenes are surfactant-like molecules, unlike the previously hypothesized concept that maltenes are the only surfactant-like component in petroleum.

INTRODUCTION

Asphaltenes are the heavy ended fraction of petroleum. They can be derived from the crude or the vacuum residue via solvent extraction[1]. It is generally believed that asphaltenes hinder refining yields, due to the formation of colloidal-like particles. In the past, a popular concept about the role of asphaltene in this colloidal complex fluid was that asphaltene form a solid-like core, coated with maltene. Here the maltenes serve as a surface active agent, separating asphaltene from the lighter material[2-5]. Elemental analyses show that asphaltenes have a H/C ratio close to unity, indicating that asphaltenes contain a substantial amount of polynuclear aromatic[6]. Additionally, heteroatoms, like nitrogen, sulfur and nickel etc., are also found in asphaltene. Based on this composition, it is natural to suspect that asphaltene molecules may also be surface active in organic solvents. If this is true, the hypothetical picture proposed previously for the asphaltene colloids may not necessarily hold.

In order to address this critical question, we performed two experiments. The first experiment was to measure the surface tension of asphaltenes in a solvent which has a surface tension higher than one end of the asphaltene molecule, while lower than the other end. The reason of selecting such a solvent was that the surface tension variation, after addition of asphaltenes, can be detected by the surface tensiometer[7]. Since the solvent surface tension lies between two ends of the asphaltene molecules, asphaltene would behave like a surfactant molecule, provided it is thermodynamically surfactant-like. If asphaltenes are indeed surfactant-like, then, maltene is not needed in the formation of colloid-like particles. The second experiment was to examine the thermodynamic reversibility of asphaltene micelles (if asphaltene micelles do form) in terms of the micellization and dissociation processes. If they are thermodynamically reversible, then, it is reasonable to name asphaltenes as petroleum surfactants. Besides the reversibility study, the kinetics of the micellization and dissociation processes was also investigated. Since the kinetics is related to the structural complexity, the structural compatibility can be qualitatively correlated with the micellization/dissociation kinetics.

EXPERIMENTAL

The asphaltenes used in this study were derived from Ratawi (Neutral Zone) vacuum residue. The residue from vacuum distillation tower (1000$^+$ F) was dissolved in heptane with 1 gm to 40 cm^3 ratio respectively, and stirred overnight. The insoluble fraction, asphaltenes, were dried under nitrogen. In order to perform the surface tension measurement, the asphaltene solids were then redissolved in nitrobenzene and pyridine respectively for surface tension measurements. The surface tension was measured as a function of concentration from 0 to 1 wt %. The surface tensiometer used was a Krüss surface tensiometer, model K10T, using the Wilhelmy plate method. The operating range was from 0 to 0.2 Nm^{-1}, with an accuracy of $\pm 1 \times 10^{-4}$ N m^{-1}. The temperature for the concentration dependent measurement was maintained at 25 °C by a water circulating temperature controller.

The second experiment involved two steps. First, 0.01% and 0.5% asphaltene/pyridine solutions were sealed in a pressure cell, and heated to 250 °C for 4 hours. The intention of this step was to dissociate the micelles. After four hours, the cell was quenched to room temperature, where the surface tension was measured as a function of time. If the samples degraded at 250 °C, the asphaltene molecular configuration would be expected to change, leading to the variation of the surface sublayer concentration, and thus the surface tension. Thus, the degradation, if it occurs, can be detected by the surface tension after quenching. In this experiment, no degradation was observed.

RESULTS AND DISCUSSION

Figure 1 and Figure 2 show the surface tension for Ratawi asphaltenes as a function of concentration with the concentration plotted in logarithmic scale, according to the Gibb's surface isotherm equation. Obviously, the surface tension exhibits a discontinuity for both cases, as the concentration exceeds a certain threshold. This is similar to micellization process in a surfactant solution. These results clearly indicate that asphaltene molecules behave like surfactants in both nitrobenzene and in pyridine. However, for most surfactant solutions, the micellization and micellar dissociation are thermodynamically reversible, it is therefore necessary to check if the asphaltene micelles are also thermodynamically reversible. The second experiment we performed was designed for this purpose.

Figure 3 and Figure 4 show the dynamic surface tensions for 0.5 % and 0.01 % asphaltenes in pyridine, after heating for four hours at 250 °C, and quenched to room temperature. Apparently, the surface tension increases slowly as a function time, indicating a gradual asphaltenes desorption, from the surface sublayer. If the systems are reversible, the equilibrium surface tension should coincide with the value obtained from the concentration measurements. From the data we obtained, the equilibrium surface tensions are 35.8 and 36.5 dyne/cm, consistent with the concentration measurements. This directly proves that the systems did not degrade and are thermodynamically reversible.

An interesting phenomenon observed in the dynamic surface tension measurements was that the surface tension increases slowly and reaches equilibrium in about 2 to three hours. This differs from typical surfactant solutions, which reach equilibrium in seconds. This delay in micellization and dissociation for asphaltene solutions is obviously due to the complex molecular structure, which requires a longer time to "pack" into a "proper" micelle, so that the system can reach its minimum free energy state. In addition to the packing kinetics, asphaltenes contain a substantial amount of polynuclear aromatic, which extend into a relatively large two dimensional area, and thus the adsorption area. As a result, it requires more energy for asphaltene desorption from the surface (i.e., higher activation energy). This means the desorption kinetics is expected to be slower than most of the surfactant solutions as well.

In order to extract the kinetic information from the dynamic surface tension measurements, we adopted a simple relation,

$$\gamma(t) = \gamma_{\infty}[1 - e^{-Kt}] \quad \text{(1)}$$

where γ_{∞} is the asymptotic surface tension, and k represents the characteristic rate for the system from the initial state to

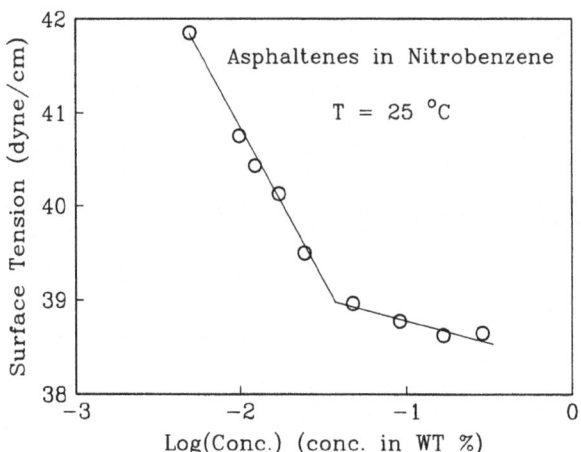

Figure 1. Surface tension as a function of asphaltene concentration in nitrobenzene. The discontinuous point represents the critical micelle concentration (CMC).

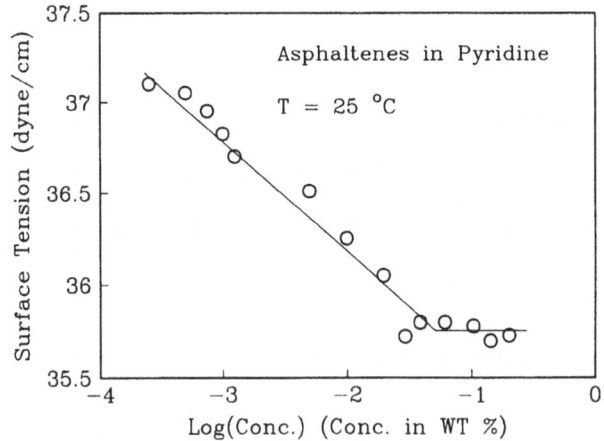

Figure 2. Same plot for asphaltene in pyridine. The CMC is at the order of 0.05 wt %.

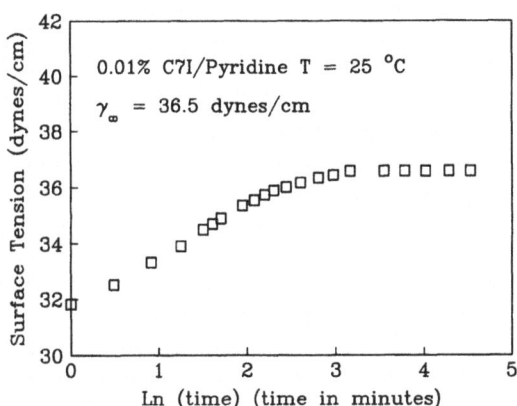

Figure 3. Surface tension as a function of time for 0.01 wt%
(below CMC) asphaltenes in pyridine, after quenching
from 250 °C. The slowly rising surface tension
indicates the slow desorption kinetics.

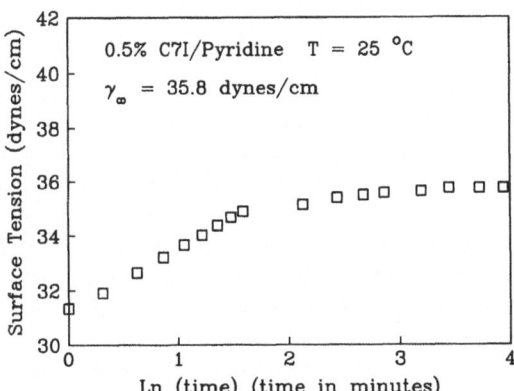

Figure 4. Same plot for 0.5 wt % concentration (above CMC).
The surface tension behavior is similar to the 0.01%
case.

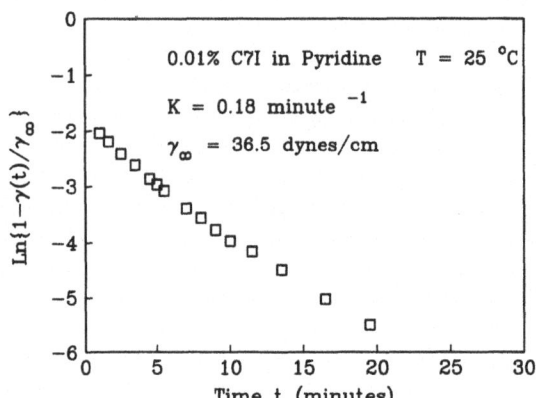

Figure 5. Analysis of the dynamic surface tension data, using
diffusion model. The single decay marks the
desorption kinetics with constant at 0.18 minutes^{-1}.

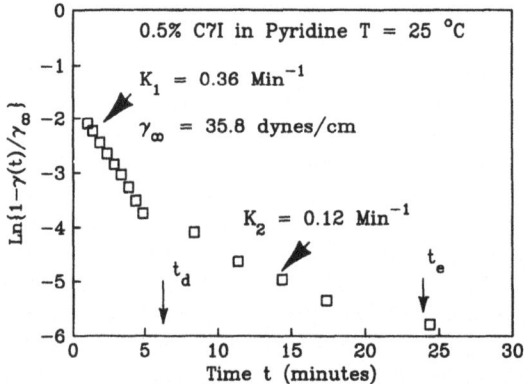

Figure 6. Same plot for the 0.5% case. Two decays are
exhibited; the first decay represents the desorption
process with a 0.36 minutes-1 kinetic constant (much
faster than 0.01% case) and the second decay shows
the micellization kinetics, which is a much slower
process, compared to the desorption process.

the new equilibrium state. This relation is will be adequate, if the desorption process and the subsequent micellization process (for 0.01% only desorption process is involved, while for 0.5% both processes are involved) can be kinetically separated. This is to say that the micellization process will not occur until the desorption is nearly completed. Should these two processes occur simultaneously, Eq.(1) will not be adequate to describe the dynamic surface tension data. For the asphaltene systems we dealt with, it appears to be adequate.

Figure 5 and Figure 6 show the results of the analyses using Eq.(1). As one can see, the 0.01% exhibits only one slope, signifying the desorption kinetics. The kinetics constant k was found to be 0.18 minutes^{-1}. On the other hand, the 0.5% case showed two distinctive slopes. The initial slope apparently represents the desorption kinetics with 0.36 minutes^{-1} kinetics constant, much faster than the 0.01% case. This faster rate is due to concentration, because the higher initial sublayer concentration (lower initial surface tension) lowers the adsorption energy, and thus the activation energy, which results in faster "reaction" (desorption) kinetics. The second slope observed corresponds to the micellization kinetics. The kinetics constant was found to be about 0.12 minute^{-1}, a very slow micellization process. This slow process clearly indicates that the asphaltene structure is truly very complex. It also suggests that any asphaltene study should take into account the effect of kinetics. In particular, when temperature, concentration, or pressure is varied, the system requires a long period of time to reach the new equilibrium state.

CONCLUSION

In conclusion, we have demonstrated that asphaltenes derived from Ratawi crude behave like surfactants in nitrobenzene and pyridine. This suggests that the classical picture of petroleum colloids, consisting of an asphaltenic core with a maltene coated surface, is not necessary true. The asphaltene micelles are thermodynamically stable and reversible upon heating and quenching. Finally, the surface desorption and micellization kinetics for asphaltenes are slow, due to inherent complexity.

REFERENCES

1.   Sheu, E. Y., De Tar, M. M. and Storm, D. A.; Solution Properties of Colloids Formed by Petroleum Vacuum Residue; Macromolecular Reports, A28, 159-175, 1991.

2.   Mack, C., Phys. Chem. 36, 2901, 1932.

3.   Pfeiffer, J. P. and Saal, R. N. J., J. Phys. Chem., 44, 1159, 1940.

4.   Eiler, H., J. Phys. Chem., 53, 1195, 1949.

5.   Ray, B. R., Witherspoon, P. A. and Grim, R. E., J. Phys. Chem., 61, 1296, 1957.

6. Speight, J., Wernick, D.L., Gould, K. A., Overfield, R. E., Rao, B. M. L. and Savage, D. W., Revue de L'institut Fracancis de Petrole, <u>40</u>, 51, Janvier-Fevrier, 1985.

7. Rosen, M. J., Surfactant and Interfacial Phenomena, John Wiley & Sons, New York, pp207, 1978.

ROLE OF ASPHALTENES IN RECOVERING HEAVY OIL THROUGH MICRO-BUBBLE GENERATION

M.R. Islam and A. Chakma*

Dept. of Geological Engineering
South Dakota School of Mines and Technology
Rapid City, South Dakota 57701 (USA)

*Dept of Chemical and Petroleum Engineering
University of Calgary
Calgary, Alberta, T2N 1N4 (Canada)

ABSTRACT

A new mechanism of fluid flow in solution-gas drive heavy oil reservoirs has been identified through experimental studies. This paper presents experimental results in order to verify previous hypothesis on fluid flow in heavy oil reservoirs in Canada. It has been postulated that solution-gas drive in these reservoirs involves simultaneous flow of gas and oil. However, gas remains in tiny droplets under reservoir conditions. A new mathematical model was proposed in order to describe peculiar pressure-dependent multi-phase flow properties. Results present experimental validation of some of the hypotheses offered by Smith[1]. More than 50 experimental tests have been performed both in a capillary tube and in a core packed with unconsolidated reservoir fluids and sands. The effects of decreasing bubble size on total fluid production and pressure drop across a pipe have been observed. Experiments with live oil and reservoir sands enable one to quantify the contribution of asphaltenes to generate bubble flow in a solution gas drive process involved in heavy oil reservoirs. Some of these results help explaining the anomalies observed in heavy oil reservoirs in Canada. Finally, field-scale reservoir simulation runs were conducted to show the effect of micro-bubble formation in Canadian heavy oil reservoirs.

Asphaltene Particles in Fossil Fuel Exploration, Recovery, Refining, and Production Processes, Edited by M.K. Sharma and T.F. Yen, Plenum Press, New York, 1994

123

INTRODUCTION

Elkins *et al.*[2] hypothesized the existence of 'worm hole' porosity in the unconsolidated sands held together only by the viscous oil. This hypothesis has often been supported by sudden collapse in injection schemes or failure of drilling and workover operations in heavy oil reservoirs of the Lloydminster area of Canada. This concept has often been used to introduce a negative skin factor in modeling heavy oil reservoirs. While this technique has given satisfactory results in several occasions, the technique has been criticized to be scientifically inaccurate. In fact, Islam and George[3] have demonstrated through laboratory experimentation that sand redistribution or even sand/fines removal would actually decrease the near wellbore permeability.

Smith[1] suggested that solution/gas drive in Canadian heavy oil reservoirs involves simultaneous flow of many oil and tiny gas bubbles. He proposed a new model for predicting performance of heavy oil reservoirs in Canada.

Kennedy and Olson[4] studied bubbles of methane in kerosene in the presence of silica and calcite crystals. They observed that bubbles formed on silica or calcite surfaces rather than in the oil medium itself. They found that the number of bubbles formed for a given volume of reservoir rock depends on the rate of diffusion of gas through oil and on the rate of pressure declined in the reservoir. One of their important observations was that the variation in gas distribution may lead to different relative permeabilities for the same gas saturation in the reservoir.

In an effort to explain discrepancy between laboratory and field observations, Stewart *et al.*[5] conducted a series of solution/gas drive tests in limestone cores. They observed that the oil recovery depends directly on the number of bubbles produced. The number of bubbles, in turn, depends on the rate of decline of pressure. Since, usually laboratory tests are performed at a pressure decline which is much higher than that in the field, laboratory tests consistently lead to higher oil recovery than do field tests.

Hunt and Berry[6] presented both experimental and theoretical studies in order to determine the probability function of bubble nucleation time. They found that the mean rate of bubble formation increases rapidly with increasing supersaturation. They presented a theory for predicting bubble size as well as bubble concentration for a given rate of pressure decline.

Wieland and Kennedy[7] re-confirmed some of the previous observations. They found that a definite supersaturation ranging from 14 to 25 psi can be imposed without forming any bubble. This range depended on the type of rock and fluid used. They attributed this variation of bubble frequency to surface tension and surface areas of the reservoir rock.

Dumoré[8] investigated the possibility of bubble
generation and the upward migration of gas and their
dependence on the rate of pressure decline. He observed
higher supersaturation pressure as the rate of pressure
decline was increased. He identified dispersion and non-
dispersion conditions depending on the nature of the porous
medium. He found through visual observation that the
solution gas-drive zone is occupied by disconnected
agglomerations of gas bubbles whereas, for nondispersion
conditions, the free gas causes a network of gas channels.
This led to much higher free-gas saturation under dispersion
conditions than under non-dispersion conditions. For a
series of consolidated core tests under different bubble
pressures, he observed that the mean free gas increases with
increasing bubble-point pressure for a given pressure drop
over bubble-point pressure ratio.

As pointed out by Smith, even though Gibbs' stability
theory indicates that bubbles larger than 200 micron size
are to grow unstably, Ward et al.[9] concluded through
experimental and theoretical study that micro-bubbles can be
in stable equilibrium under the constraint of a closed
volume and for reasonable conditions of liquid temperature
and pressure. The analysis of Ward indicates that it is
possible to generate micro-sized bubbles in a porous medium.
These bubbles being much smaller than the average pore
throat size, gas is not restricted to flowing only as a
continuous phase as postulated in previous reservoir flow
calculations. In this regard, the analysis of Ward and
Levart[10] should be mentioned. Their analysis indicates that
very small bubbles may be present in thermodynamically
stable form in the presence of a rough surface. This surface
may be either that of the boundary or that of particles
suspended in the solution. Similar surface roughness maybe
provided by asphaltene molecules which are usually found in
abundance in heavy oil. Micro-bubbles formed in this way are
stable in such a way that only further reduction in pressure
may make them grow, reducing the possibility of growth by
coalescence.

Recently Islam[11] has reviewed the role of asphaltenes
in recovering heavy oil. Even though asphaltenes are
attributed to creating enormous production problems during
oil productions, he recognized that the presence of
asphaltenes may lead to more bubble capture by heavy oil
leading to improved oil recovery during primary production.
This paper reports a series of experimental results in order
to investigate the effect of various parameters and role of
micro-bubbles on heavy oil recovery.

EXPERIMENTAL

Two sets of experiments were carried out in this study. They
are: (1) study of the mechanics of bubble flow in a straight
capillary tube; and (2) study of the mechanics of bubble flow
in core packed with unconsolidated sands.

Capillary tubes of 2mm diameters and 2m length were used for all the capillary flow runs. The tube was made of steel and could withstand an internal pressure of up to 35000 psi. Simultaneous injection of gas and liquid was carried out through the inlet. Pressure drop along the capillary tube was measured using several pressure transducers. Micro-bubbles were generated using sintered metal cap of known pore openings. The dimension of micro-bubbles related to the flow rate of gas injection and relative gas/oil viscosity. The produced fluid was collected in test tubes and the liquid volume was measured after the liquid was degasified. Gas volume injected was measured with a mass flow meter mounted at the entrance of the gas stream.

The second series of experiments were conducted in a horizontal core packed with unconsolidated sands. The sand pack was under a triaxial pressure of 2100 kPa. The core was prepared by wet packing using the vibration technique. Once packed, the core was dried by blowing air overnight. A vacuum was then pulled and water imbibed. Water was then pumped through the core and an accurate mass balance was performed to obtain the pore volume of the core. Following this, an oil flood was carried out in order to establish irreducible water saturation. A series of runs were performed using simultaneous injection of oil and gas. Aberfeldy sand was used throughout this series of experiments. Both crude oil and Dow corning oil was used for this study. Methane was used as the gas phase.

RESULTS AND DISCUSSION

The first phase of the experiments were organized to determine the role of micro-bubbles in recovering heavy oil. This was followed by determination of the role of asphaltenes in increasing micro-bubble capture.

Dow corning oil of viscosities of 10, 1000, and 5000 mPa.s were used in order to determine pressure drop in a capillary tube. Micro-bubbles were injected with oil simultaneously at different gas-oil volume ratios. Figure 1 shows the pressure drop across the capillary tube. For this particular case, total flow rate was kept constant at 400 cc/hr. At gas volume fraction of 0.1, the pressure drop was 5.2 kPa. This value was somewhat higher than segregated flow of liquid and gas. However, note that this may not be the case in a porous medium for which free gas is trapped until a critical gas saturation is reached. This leads to actually increased pressure drop in the presence of segregated flow in porous media. Figure 1 also indicates a sharp decline in pressure drop as gas volume fraction was increased. Experimental runs were not continued beyond a gas volume fraction of 0.5 for which it was difficult to maintain the gas phase in micro-bubbles.

Figures 2 and 3 show pressure drop results for oil viscosity of 1000 and 5000 mPa.s, respectively. These two

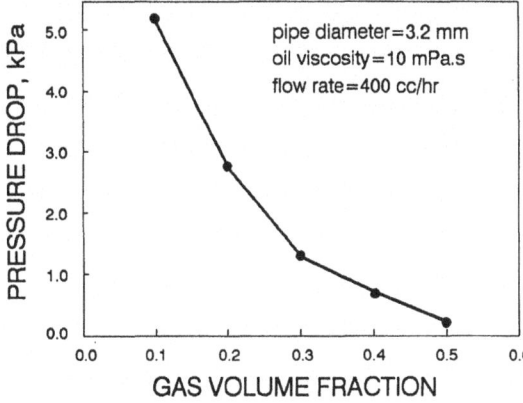

Fig. 1. Pressure drop for Dow corning oil of viscosity of 10 mPa.s

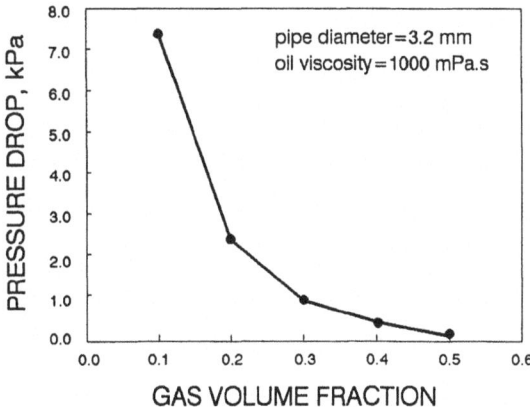

Fig. 2. Pressure drop for Dow corning oil of viscosity of 1000 mPa.s

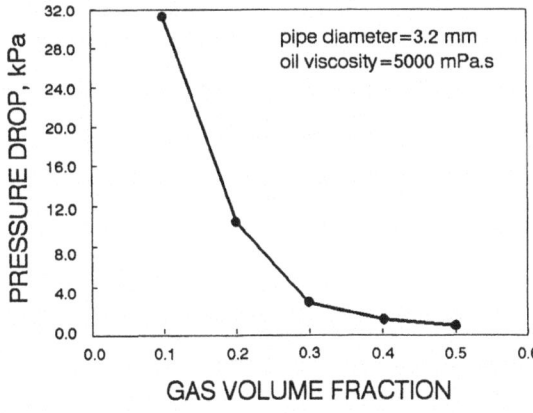

Fig. 3. Pressure drop for Dow corning oil of viscosity of 5000 mPa.s

runs were conducted at a total flow rate of 100 cc/hr. As
can be seen in these figures, pressure drops decline rapidly
with increasing gas volume fraction for each run. For all
these cases of liquid and micro-bubble flow, mixture
viscosity may be best approximated by

$$\mu_m = \mu_l^f \mu_g^{1-f}$$

where μ_m is the mixture viscosity, μ_l is the liquid
viscosity, μ_g is the gas viscosity, and f is the liquid
volume fraction. This expression was originally given by
Arrhenius and was modified by Hagerdorn and Brown[12]. The
deviation of observed apparent viscosity was the greatest
when gas volume fraction was higher.

Another significant finding of this experimentation is
the existence of micro-bubbles at high gas saturations. Dow
corning oil was used for these runs. This oil does not
contain any solid particle which might act as nucleation
sites for micro-bubbles. The formation of micro-bubbles was
watched under microscope through the transparent window of
the capillary tube. The experiment was conducted at a mean
pressure of 600 kPa. Microphotographs were taken as snap
shots in order to monitor the velocity of micro-bubbles.
Surprisingly, average velocity of micro-bubbles was 1.2
times higher than the liquid velocity. Implication of such a
behavior is an increase in fluid flow when the gas phase is
in the form of micro-bubbles as opposed to a continuous gas
phase.

In order to determine the role of asphaltene particles
as nucleation sites for micro-bubbles, a series of runs were
conducted with crude oil which contained asphaltenes. Figure
4 shows pressure drop as a function of gas volume fraction
for the crude oil. Heavy crude of dead oil viscosity of 750
mPa.s was used for these tests. Micro-bubbles were not
visible for crude oil which is essentially opaque and does
not allow any microphotography of the bubbles. The crude oil
contains some amount of asphaltene which may be an excellent
nucleation site for micro-bubbles. The presence of
asphaltene appears to ease the flow of the mixture at lower
values of gas volume fractions. However, as gas volume
fraction increases, pressure drop is actually higher as
compare to that of Dow corning oil. This behavior in a
capillary tube will translate into poor recovery at later
stages of oil recovery in heavy oil reservoirs.

The effect of micro-bubble size is shown in Figure 5.
Even though the shapes of the pressure drop versus gas
volume fraction curves are similar for all different bubble
sizes, actual pressure drop value tends to increase as the
bubble size decreases. This indicates that the generation of

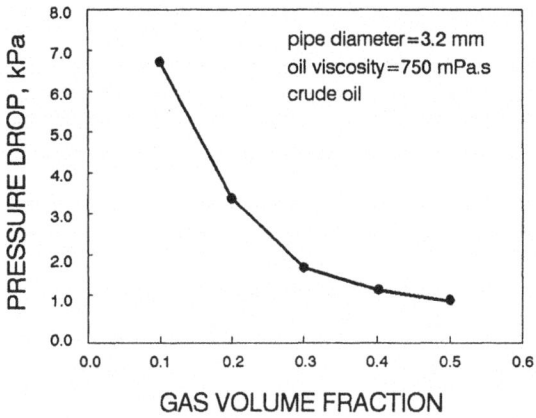

Fig. 4. Pressure drop for crude oil

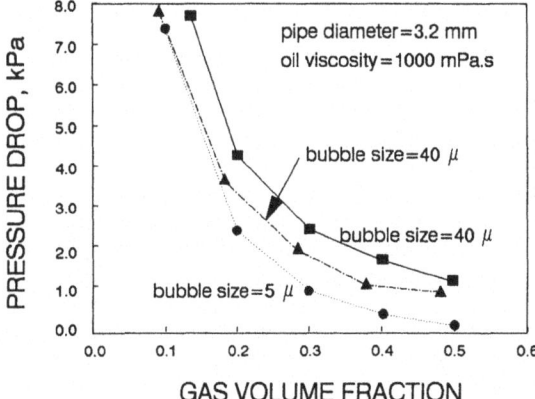

Fig. 5. Effect of bubble size

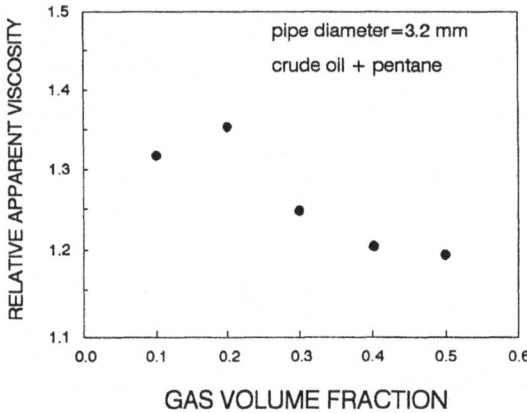

Fig. 6. Relative apparent viscosity for crude oil and pentane

tiny micro-bubbles in heavy oil reservoirs will enhance fluid flow for a given pressure decline. Actual size of micro-bubbles in heavy oil reservoirs is not known. However, the results in Figure 5 indicate that a substantial increase in fluid flow may occur by mere decrease in gas bubble size.

To investigate the contribution of asphaltene in solution, a few runs were conducted by adding pentane to the crude oil. This addition leads to precipitation of asphaltene and thereby takes away the advantage of having asphaltene as nucleation sites. As a consequence, the flow of gas and liquid is restricted. Since the addition of pentane decreased the bulk viscosity of the liquid phase, results had to be compared as relative apparent viscosity. Figure 6 shows relative apparent viscosities as a function of gas volume fraction when crude oil was injected along with micro-bubbles after precipitation of asphaltene particles.

A series of core flood tests were carried out to study bubble flow behavior in the presence of heavy crude oil. Figure 7 compares the effect of micro-bubbles with that of continuous gas/oil injection. Both these runs were performed at a constant oil/gas ratio of 4. It can be seen from Figure 7 that, by creating micro-bubbles, the pressure drop across the core is considerably reduced. As was pointed out earlier, the free gas (in the form of large interconnecting bubbles) gets easily trapped in the porous medium leading to increased pressure drop. As the steady state is reached, pressure drop in the case of continuous gas flow is twice as high as that observed in the presence of micro-bubbles, even though S_o/S_g remained the same for both cases. This figure explains much of the flow behavior which produces lot more oil than that expected from conventional relative permeability analysis. Due to low diffusion of gas in heavy oil, often non-equilibrium phenomena occur. Such a behavior is also noticed in both gas dissolution or bubble generation.

In order to quantify micro-bubble release in the presence of asphaltene, results were compared with Dow corning (no asphaltene) and crude oil and pentane solution (20:1 ratio). Viscosity was the same for crude oil and Dow corning oil. However, crude oil and pentane solution had a much lower viscosity than that of the other two oils. Micro-bubbles and oil were injected in a core simultaneously until steady state was achieved. The core holder was maintained at a pressure of 1700 kPa and was left at that pressure for three days. Initial gas/oil ratio was 1 at standard conditions. After three days, high pressure samples were collected and was suddenly exposed to atmospheric pressure. The amount of unescaped fluid volume was measured with time. These values are reported in Figure 8. Note that both Dow corning oil and crude oil and pentane solution released bubbles at approximately the same rate. However, the rate of release is much slower for crude oil which, after an initial sudden release, releases gas at approximately constant rate. This rate is much slower than

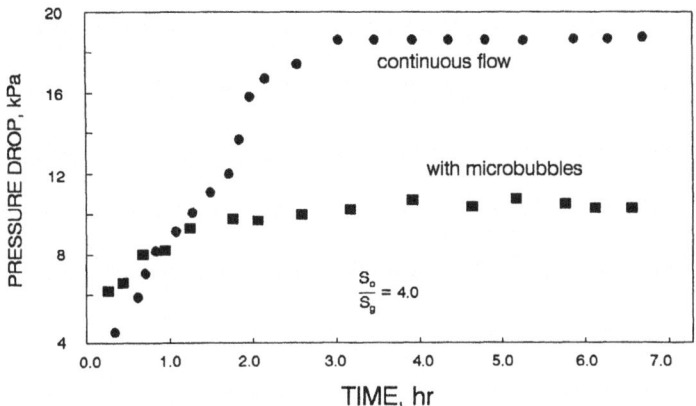

Fig. 7. Pressure drop in a consolidated core

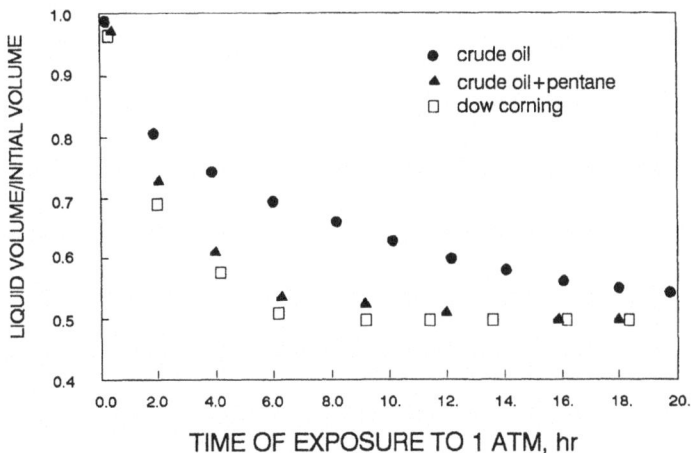

Fig. 8. Delay in bubble release

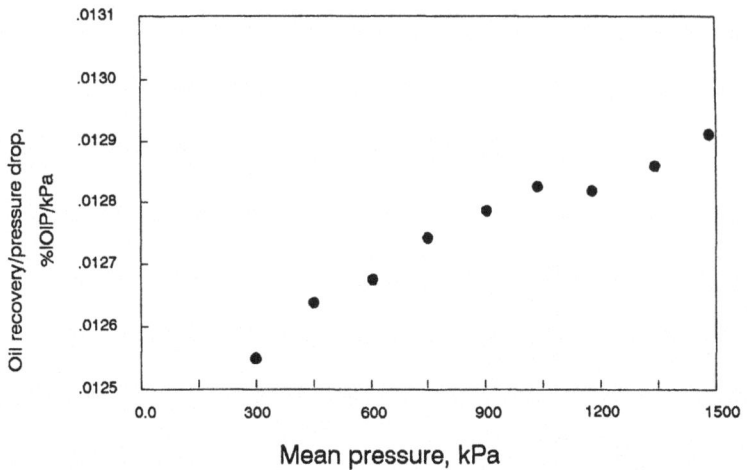

Fig. 9. Effect of mean pressure on oil recovery efficiency

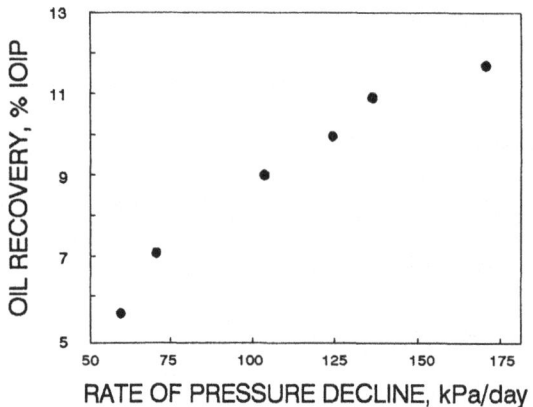

Fig. 10. Effect of pressure decline rate on oil recovery

Fig. 11. Effect of pressure decline rate on relative gas oil volume

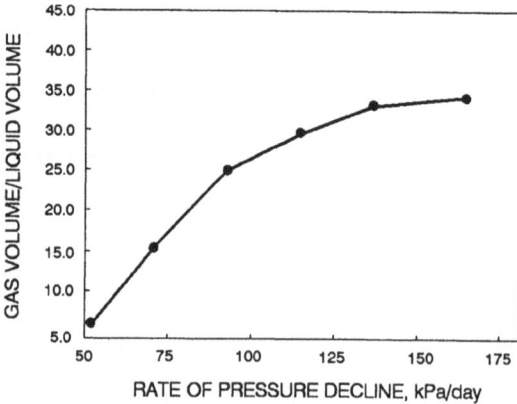

that of the other two types of oils. As indicated earlier, the presence of asphaltenes may provoke higher amounts of micro-bubble capture and lead to a delayed bubble release.

The effect of confinement pressure on bubble size has been studied by several authors. Such an investigation in porous media is very difficult. Consequently, the effect of mean pressure on the recovery efficiency of heavy oil (expressed as IOIP recovered/kPa) was investigated. The effect of mean pressure is shown in Figure 9. Very clearly, as the mean pressure is increased, recovery efficiency goes up. Our investigation continued only up to a pressure of 1500 kPa. For that range of pressure, there is no apparent leveling off observed. This increase in recovery efficiency may be due to the generation of very small bubbles as the pressure was increased.

In order to investigate the effect of pressure decline rates on oil recovery, six core flood tests were conducted with crude oil keeping all variables except pressure decline rate constant. Results of these runs are shown in Figure 10. Note that here we investigate recovery by solution gas drive alone. Consequently, ultimate oil recovery values are rather small. However, this is indeed the range of oil recovery observed under primary production in heavy oil reservoirs of Canada. As can be seen in Figure 10, the ultimate oil recovery by solution gas drive increases as the pressure decline rate increases. As the pressure decline rate is increased the number of bubbles is increased leading to finer bubbles. This is responsible for increased oil recovery. This trend can be observed also when gas liquid volume ratios are reported as a function of pressure decline rate. This is shown in Figure 11. Even though gas bubble radii were not monitored in these runs, Figure 11 clearly indicates that the bubble size decreases with increasing pressure decline rates.

Asphaltene may play an important role in capturing micro-bubbles and the bubble size may depend on the concentration of nucleation sites provided by asphaltene molecules. The effect of asphaltene concentration on gas/liquid volume is depicted in Figure 12. There appears to be a distinct impact of asphaltene concentration on gas-bearing capability of the mixture. This translates itself as one of the major factors in determining the bubble size.

Two tests were carried out in order to investigate the effectiveness of micro-bubbles as displacing agent. The coreflood tests were conducted using dead crude oil. Gas flood was carried out by injecting micro-bubbles in one case and continuous gas stream in another. The results are plotted in Figure 13. This figure shows that micro-bubbles indeed increase recovery by a substantial amount. Generation of micro-bubbles during a field scale gas injection scheme is not a trivial task. However, in many cases, micro-bubbles may be present in the reservoir. This would enhance oil production from reservoirs.

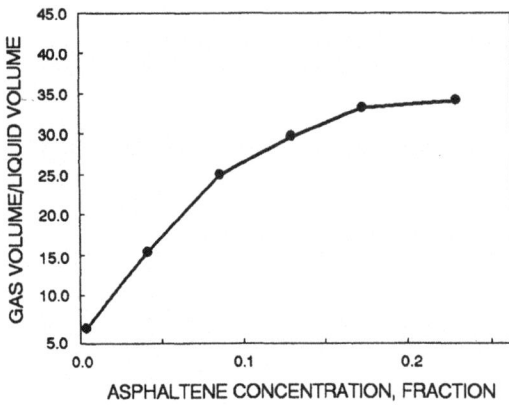

Fig. 12. Effect of asphaltene concentration on relative gas oil volume

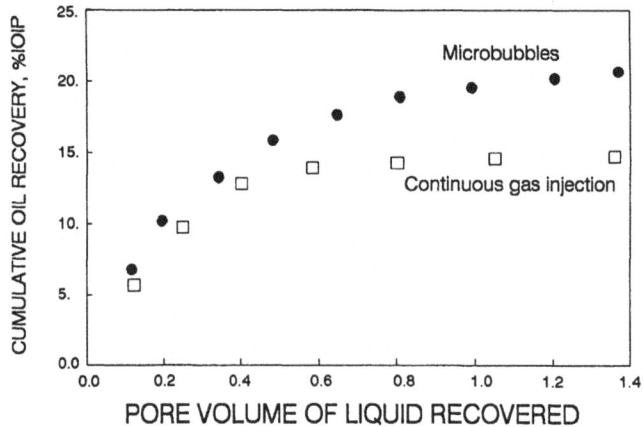

Fig. 13. Role of micro-bubbles as displacing agent

Fig. 14. Gas-oil relative permeability curves for a solution-gas drive process

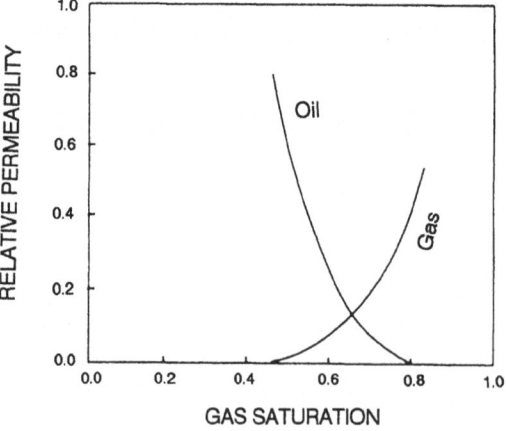

NUMERICAL

Describing solution gas drive in heavy oil reservoirs has been a research material for some time. Smith[1] proposed that oil and gas phases in heavy oil reservoirs be treated as a single phase with variable viscosity and density depending on gas concentration and pressure. This technique may be appropriate when micro-bubbles are present throughout the production history of the field. However, if gas saturation is very high at the later stages of the oil recovery, we found it more appropriate to use two-phase flow with relative permeability curves carefully crafted for the system in question. Our analysis indicates that, for heavy oil reservoirs, gas/oil relative permeability curves should be similar to what are shown in Figure 14. These curves are only valid for solution gas drive and should not be used in the case of gas injection or waterflood (three-phase flow). Note that the relative permeability curves are such that there are only single phase flow at initial stages of the production history. However, once gas is mobilized as a segregated phase the relative permeability to gas increases rapidly.

Table 1 lists reservoir and fluid parameters used for numerical simulation. A numerical simulator developed earlier[13] was used for simulating the heavy oil reservoir. The original simulator is capable of modeling rigrous wellbore phenomena. However, this option was inactivated since such phenomena are deemed unimportant for a vertical

TABLE 1. RESERVOIR AND FLUID PROPERTIES

PARAMETERS	DATA
INITIAL OIL SATURATION	80%
INITIAL WATER SATURATION	20%
PERMEABILITY	4 μm
OIL VISCOSITY	1000 mPa.s
INITIAL RESERVOIR PRESSURE	8300 kPa
POROSITY	32%
WELLBORE DIAMETER	8.9 cm
AREA SIMULATED	250m x 250m
WELL TYPE	SINGLE VERTICAL WELL

wellbore. Numerical simulation runs were conducted for a time period of twenty years. Three different numerical simulation approaches were taken. They are as follows:

(1) single-phase treatment of the heavy oil as proposed by Smith[1].

(2) conventional two-phase treatment of the heavy oil/gas system.

(3) single-phase treatment of the heavy oil (following Smith[1]) with delayed bubble release from the oil, followed by two-phase flow of the heavy oil and gas system.

During single-phase treatment of the heavy oil (Approach 1), it was assumed that the viscosity of the mixture is given by the Equation 1. The amount of gas present in the heavy oil was dictated by PVT data which were taken from Islam *et al.*[14] For Approach 2, numerical simulation was conducted following conventional reservoir simulation approach, that is, by using laboratory determined two-phase relative permeability curves. The aqueous phase was considered to be immobile, consequently only gas-oil relative permeability curves were used. These curves were taken from Islam *et al.*[14], as used for unstable displacement. For the Approach 3, delayed bubble release, i.e., the delayed initiation of two-phase flow was modeled by using a different set of relative permeability, as shown in Figure 14. During initial single-phase flow, a mixture of heavy oil and micro-bubbles was assumed to flow following Equation 1. Figure 15 shows oil flow rate and cumulative oil recovery prediction using Approach 3 for the system considered. Note that the oil production of around 15% of the initial oil place from such a heavy oil reservoir is typical of Canadian reservoirs. Figure 16 compares the cumulative recovery curves obtained by using different approaches. Note that the single-phase assumption (Approach 1) gives similar results as Approach 3 during initial stages of oil production.

CONCLUSIONS

Several mechanisms of fluid flow in solution-gas drive heavy oil reservoirs are studied. The hypothesis of micro-bubble formation in the presence of Canadian heavy oil is supported by experimental results. Possible consequences of micro-bubble formation and its impact on oil recovery are reported. Also studied is the effect of mean pressure as well as asphaltene concentration on bubble flow. Asphaltenes and high confinement pressure are found to contribute to the formation of very small gas bubbles and higher oil recovery. Micro-bubbles are found to have much higher recovery efficiency over continuous gas streams. A set of relative permeability curves is presented to study solution/gas drive in the presence of micro-bubbles. Field

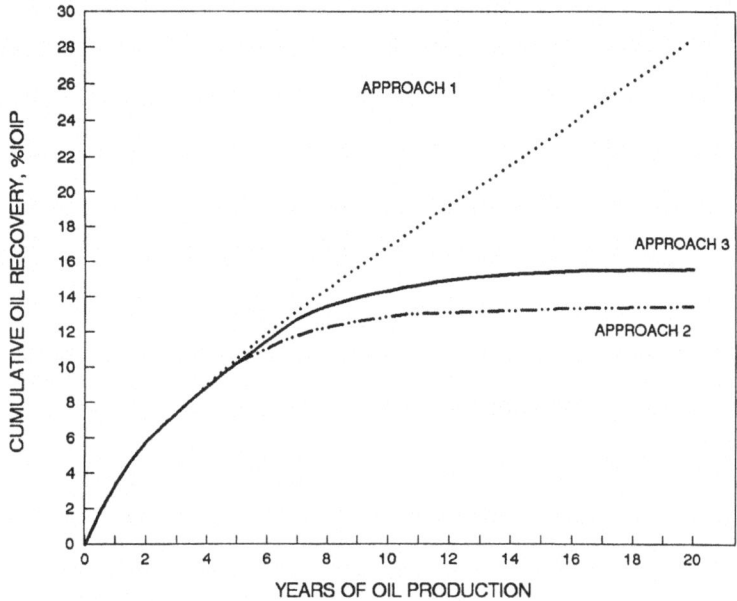

Fig. 15. Oil production rate and cumulative production using Approach 3

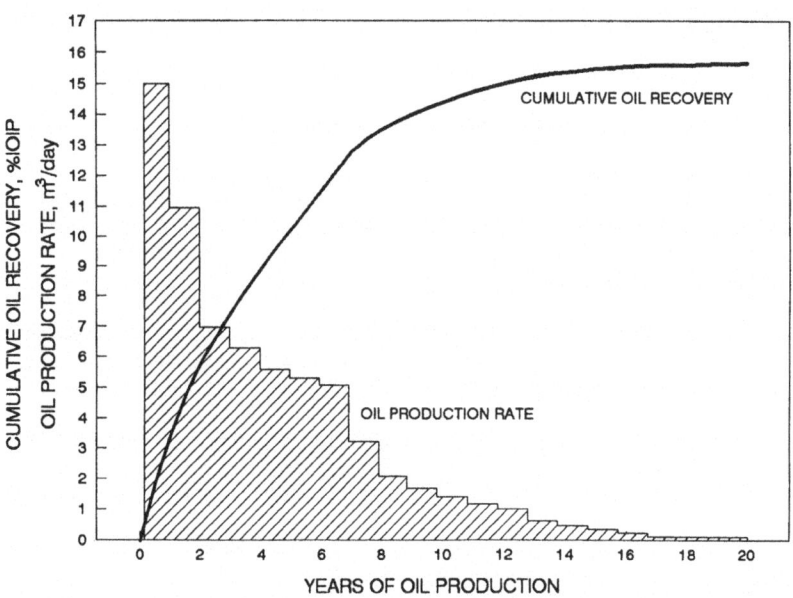

Fig. 16. Comparison of different simulation approaches

scale numerical simulation is conducted to demonstrate the most effective way to describe fluid flow in a heavy oil reservoir.

REFERENCES

1. Smith, G. E., "Fluid Flow and Sand Production in Heavy Oil Reservoirs Under Solution-Gas Drive", SPE Production Engineering , vol.3; 169-179, May,1988.

2. Elkins, L.F., Morton, D. and Blackwell, W. A., "Experimental Fireflood in a Very Viscous Oil-Saturated Sand Reservoir, S. E. Pauls Valley Field, Oklahoma", SPE paper no. 4086, presented at the 47th Annual Fall Meeting, Texas, 1972.

3. Islam, M. R. and George, A. E., "Sand Control in Horizontal Wells in Heavy Oil Reservoirs", Journal of Petroleum Technology, vol.43; 844-853, July,1991.

4. Kennedy, H. T. and Olson, C. R., "Bubble Formation in Supersaturated Hydrocarbon Mixtures", Trans. AIME, vol. 195, 271-278, 1952.

5. Stewart, C. R., Hunt, E. B., Jr.,Schneider, P. N., Geffen, T.M., and Berry, V.J., Jr., "The Role of Bubble Formation in Oil Recovery by Solution Gas Drives in Limestones", Trans. AIME, vol. 201, 294-301, 1954.

6. Hunt, E. B. and Berry, V. J., Jr.,"Evolution of Gas From Liquids Flowing Through Porous Media", AIChE J,vol.2(4); 560-567, Dec.,1956.

7. Wieland, D.R. and Kennedy, H.T.,"Measurements of Bubble Frequency in Cores", Trans. AIME, vol. 210, 122-125, 1957.

8. Dumoré, J.M., "Development of Gas-Saturation During Solution-Gas Drive in an Oil Layer Below a Gas Cap", Society of Petroleum Engineers Journal, September, vol. 10; 211-218, 1970.

9. Ward, C.A., Tikuisis, P. and Venter, R. D., "Stability Analysis in a Closed Volume of Liquid-Gas Solution", J. Appl. Phys., vol. 53 (9), 6076-6084, 1982.

10. Ward, C. A. and Levart, E., "Conditions for Stability of Bubble Nuclei in Solid Surfaces Contacting a Liquid-Gas Solution", J. Appl. Phys., vol. 56 (2) (1984), 491-500.

11. Islam, M.R., "Role of Asphaltenes in Heavy Oil Recovery", in "Asphaltic Materials", Ed. T.F. Yen, Elsevier Publication,(IN PRESS).

12. Hagerdorn, A.R. and Brown, K.E.,"Experimental Study of Pressure Gradients Occurring During Continuous Two-Phase

Flow in Small Diameter Vertical Conduits", Journal of Petroleum Technology, vol. 17; 475-484, April,1965.

13. Islam, M.R. and Chakma, A., "Comprehensive Physical and Numerical Modeling of a Horizontal Well", paper SPE 20627 presented at the SE Annual Technical Conference and Exhibition, New Orleans, LA, pp.111-123, 1990.

14. Islam, M.R., Erno, B.P., and Davis, D., "Hot Gas and Waterflood Equivalence of In Situ Combustion",Journal of Canadian Petroleum Technology, vol.31(8); October, 1992.

THE EFFECT OF SURFACTANTS TO THE CONVERSION OF COAL AND COAL-DERIVED ASPHALTENES IN COAL LIQUEFACTION PROCESS

Jiunn-Ren Lin, Teh Fu Yen and George Hsu

Civil and Environmental Engineering
University of Southern California
Los Angeles, California
U.S.A.

ABSTRACT

Due to the agglomeration of coal particles and coal-derived asphaltenes, conventional processes for coal liquefaction require its reactions under severe conditions. Even at a high temperature and a high hydrogen pressure, the yield of coal liquefaction is still not satisfactory. In this investigation, surfactant is introduced in the coal liquefaction process to reduce the agglomeration of the coal particles and coal-derived asphaltenes. Thus, the yield of coal liquefaction is improved significantly. In the surfactant screening test, six selected surfactants were examined at elevated temperatures separately in both coal/coal solvent systems. The viscosity of coal suspension and the suspended coal concentration are used as indicators to determine the performance of surfactants. The results show that surfactant calcium lignosulfonate has the best performance among selected surfactants which may be due to the increase of surface charges and double layer expansion. In the autoclave experiments, the suggested surfactant not only raises the yield of coal liquefaction but also changes the distribution of products. Under the same condition, almost twice the yield was obtained for both coal and lignite samples with the aid of surfactants.

Asphaltene Particles in Fossil Fuel Exploration, Recovery, Refining, and Production Processes, Edited by M.K. Sharma and T.F. Yen, Plenum Press, New York, 1994

141

INTRODUCTION

Coal is a very important energy resource in the world. Due to its storage and transportation difficulties, the usage of coal is not as convenient as the usage of petroleum products. To overcome these drawbacks, coal liquefaction process is used to convert coal from solid phase into liquid phase.

This pyrolysis process can be employed to convert coal into char or coke, coal tar, and gas. When coal is gradually heated, bond-breaking reactions within the network structure begin at 350-400 °C. During this stage, coal is converted into some volatile fractions and a complex condensable liquid, which is called coal tar. Chemical changes occur in the remaining solid where bond-making reactions contribute to the formation of a nonvolatile char or coke at high temperature. Thus, the pyrolysis process can be employed to convert coal into either char or coal tar, and gas.[1-3] When the fluid fuels are desirable, hydrogen is added to increase the H/C ratio and the yield of coal tar. Liquefaction processes use a partially hydrogenated coal-derived liquids to serve as hydrogen source and transfer agent for degrading coal.[4]

The conversion of coal through hydropyrolysis has been suggested to progress via the following sequence:[5,6]

Coal   **Asphaltene**   **Oil**

Preasphaltene

The benzene insoluble-pyridine soluble fraction has been named "preasphaltene" and is believed to be intermediate between coal and classical asphaltene. During the process of coal liquefaction with the assistance of solvent extraction, coal is hydrogenated first into preasphaltenes or asphaltenes and then into the smaller oil products. The preasphaltenes are highly polar due to the presence of several functional groups. The asphaltene intermediates consist of high-molecular-weight polyaromatic or polycyclic nuclei with linkage of heteroatoms such as N, O, and S.[7] Due to the polar nature, coal, preasphaltenes and asphaltenes participate in charge transfers and agglomerate together as colloidal clusters, and tend to agglomerate into aggregates with high molecular weight. Consequently, high energy input, in terms of temperature, total pressure and hydrogen partial pressure, are often needed to break the agglomerated coal particles into smaller fragments for hydrogen accessibility and the subsequent hydrogenation of preasphaltenes and asphaltenes into oil.

Most of the surfactants have a structure belonging to amphiphatic structure, including both lyophobic group and lyophilic group. The addition of surfactant will significantly affect the physical properties of the system.[8,9] It has been widely used in enhanced oil recovery to enhance the dispersion of asphaltene.[10] Surfactants have been considered in the study to be effective in liquefying coal in at least two ways. One way is to disperse and extract coal species at relatively low temperatures. This may occur as a result of an interaction between the polar surfactant, acting as a Lewis base, and the acid sites of the coal structure. Another way is the prevention of the agglomeration of asphaltenes and preasphaltenes which will thereby facilitate their dissolution.

Existing coal liquefaction processes may benefit from the addition of surfactants in terms of improved product distributions and increased yields, as well as reduced operating and capital costs. Viscosity and suspended coal concentration are two indicators used for the preliminary surfactant screening. The autoclave experiments which are conducted at Jet Propulsion Laboratory (JPL) are followed to demonstrate the effect of the surfactants at high temperature (350°C) and high hydrogen pressure (1900 psi). The results show that the addition of surfactants does increase the yield significantly for both bituminous coal and lignite samples.

EXPERIMENTAL

There are two samples used in this investigation: pulverized Illinois #6 bituminous coal and Texas lignite. Both samples were obtained from Penn State Coal Bank (PSOC 1493 & PSOC 1442). The particle size of Illinois coal is less than 200 mesh and of the Texas lignite sample is less than 20 mesh. A coal-derived solvent was provided by proof-of-concept facility at Pittsburgh Energy Technology Center (PETC), Alabama. This solvent was produced from a two-stage modified solvent refined coal as a cut of the Mid Distillate product from 400 to 650°F.

SURFACTANT SCREENING: Six surfactants involving three different types (shown in Table 1) were used in the study. All the surfactants used are commercially available. No further treatment was provided before its usage.

Two indicators were used to determine the performance of surfactant in the coal/coal solvent system: viscosity and suspended coal concentration. Viscosity was measured by the Brookfield LV model viscometer with a spindle speed of 60 rpm. The viscosity for each sample was obtained from the mean of 10 runs of measurement. A Beckman UV/Visible spectrophotometer at a wavelength of 600 nm provided absorption for the coal particle suspension. A calibration curve for the relationship between the dispersed coal concentration and the reading of absorbance for Illinois bituminous coal were established prior to the experiment.

Table I. Name and types of surfactants used in the surfactant screening.

TYPE	NAME
Anionic	Calcium lignosulfonate
Anionic	Aerosol OT
Cationic	Amberlite
Cationic	Hexadecylamine
Nonionic	Tween 80
Nonionic	Alrosperse

For pulverized Illinois #6 coal sample, 10% of coal/coal solvent suspensions were prepared at room temperature. Surfactants with various amounts and/or species were added to the suspension prior to the experiment . These suspensions were heated at 100°C for 1 hour before determining its viscosity and suspended coal concentration. Spin bar agitation was used to suspend the coal particles.

After being heated for one hour, the viscosity of coal suspension is determined at 100°C. To determine the dispersed coal concentration, the suspensions remained on the table without agitation for 1 hour. The precipitation of coal particles occurred during this period. Samples were taken from one inch below the surface and then diluted 100 times with coal solvent before the UV/visible test. Through the calibration curve and dilution factor, the suspended coal concentration could be determined.

For the Texas lignite sample, the procedure was the same as Illinois coal sample but the temperature 70°C was used instead of 100°C. In addition, there were only three selected surfactants, one for each category, used in the screening: calcium lignosulfonate, Amberlite and Alrosperse.

For the effect of temperature on the surfactant performance, the viscosities of bituminous coal/coal solvent suspensions with various amounts of surfactants were measured after the suspensions were heated for 1 hour at temperatures ranging from 40 to 100°C.

AUTOCLAVE TESTING: The autoclave experiments for coal liquefaction were conducted in a 1-liter stirred autoclave reactor. Through the results of surfactant screening and preliminary studies, lignin sulfonate (including sodium lignosulfonate and calcium lignosulfonate) and/or triton x-100 were selected in this study. For each run, 100 g of coal or lignite, on an as-received basis, was added to a premix of the solvent consisting of 200 g solvent containing the surfactant, all introduced at room temperature while stirring. The resulting coal/coal solvent slurry-blend was transferred into the autoclave reactor. Hydrogen pressure initially at an average of 1000 psig was charged into the system and increased to 1900 psig at 350°C. The retention time for the reaction was one hour. After the reaction, the system was allowed to cool down by stirring.

Products from autoclave were subjected to a series of physical and solvent based separations to determine product distribution. Separation of products by distillation, solvent extraction, and volume filtration were conducted.

RESULTS AND DISCUSSION

SURFACTANT SCREENING: The presence of coal particles is the primary contributor to the viscosity of coal/coal solvent system. In the same system, the degree of coal particle aggregation will affect the volume fraction and Huggins coefficient in Einstein's equation and the viscosity of sample will increase with the increasing aggregation;[11,12] whereas, the suspended coal concentration will decrease with the increasing aggregation. According to Stoke's equation, the terminal settling speed of particles in the solution is proportional to the square root of particle diameter. Small particles are suspended easier in the solution than large particles, and the suspended coal particle concentration will decrease with increasing particle size. Thus, viscosity and suspended coal concentration have an opposite trend with the degree of aggregation.

A. Pulverized Illinois #6 bituminous coal
    The viscosity and suspended coal concentration in the coal/coal solvent system with various surfactants and/or its concentrations are shown in Figures 1 and 2. The results show that regardless of which surfactant is introduced, the viscosity of the system decreases and the suspended coal concentration increases. It also reveals that the viscosity reduction increase with surfactant concentration.

Combined with the results shown in the viscosity and suspended concentration studies, we note that the performances of selected surfactants are in the following order: calcium lignosulfonate > hexadecylamine > Aerosol OT > Alrosperse > Tween 80 > Amberlite. Even though the data show that the variation of performance within each category is larger than that among categories, on the basis of the same weight of surfactants added, the ability of enhancing the suspension of coal increases from cationic surfactants, nonionic surfactants to anionic surfactants. Thus, calcium lignosulfonate, Alrosperse and Amberlite (or hexadecylamine) each belonging to the three different categories respectively, are selected for further studies.

Most of the colloidal particles are negatively charged. The results may support the fact that the coal particle is negatively charged, too. The adsorption of anionic

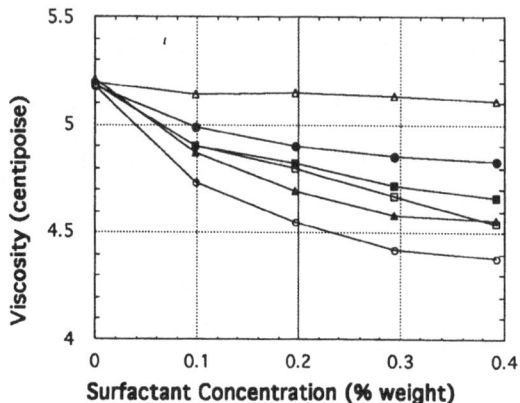

Figure 1. Viscosity of pulverized coal/coal solvent
suspensions with different contents of surfactants
(open circle: calcium lignosulfonate; solid
circle: Tween 80; open square: Aerosol OT; solid
square: Alrosperse; open triangle: Amberlite;
solid triangle: hexadecylamine).

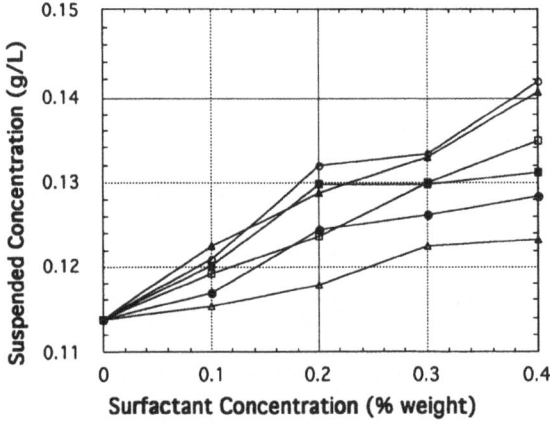

Figure 2. Coal concentration at different content of nonionic
surfactants (open circle: calcium lignosulfonate;
solid circle: Tween 80; open square: Aerosol OT;
solid square: Alrosperse; open triangle:
Amberlite; solid triangle: hexadecylamine).

surfactant on the coal surface causes the particle charge to increase. The increase of the double layer thickness around the coal particles raises the repulsive forces. The attractive forces between particles are reduced due to the increasing particle distance. Nonionic and cationic surfactants reduce the attractive force in the same way, but the cationic surfactant reduces the particle charge that reduces the repulsive force. Therefore, the performance of anionic surfactants will be better than nonionic and cationic surfactants.

The surfactant hexadecylamine has a performance similar to that of calcium lignosulfonate. It is believed that hexadecylamine adsorbs on the coal particle due to the ion pairing or ion exchanges. The solvent-particle affinity increases, and attractive forces are reduced by the long chain lyophile. Thus, its good performance follows, since the interactions among asphaltene, surfactant, solvent and water are complicated and important. Without further investigation, not much of a conclusion could be stated.

TEXAS LIGNITE: Three surfactants belonging to various types are examined in the lignite/coal solvent system. The viscosities of the coal/coal solvent suspensions with various surfactant concentrations are represented by Figure 3. It shows that all surfactants can reduce the viscosity of coal suspension, like in Illinois bituminous coal suspension. In addition, the performance of these three surfactants in lignite suspension has the same trend as in Illinois bituminous coal suspension. This may imply that the surface characteristics of these two coal samples may be similar.

TEMPERATURE EFFECT: To predict the stability of surfactant at high temperature condition, the percentage viscosity reduction at the temperature range of 40 to 100°C is measured while other variables are held constant. Table 2 and Figure 4 represent the viscosities reduction of 10% pulverized coal/coal solvent suspensions with various surfactant contents at various temperature. It is obvious that viscosity reduction increases with temperature for all surfactants. It may imply selected surfactants have better performance at higher temperature than at low temperature.

When temperature rises from 40 to 100°C, the viscosity reduction of coal/coal solvent system with anionic surfactant increases 3 folds, twice or once for the ones with nonionic surfactant and cationic surfactant. The data show that the performance of hexadecylamine is the best performance at low temperature, but its performance is not as good as calcium lignosulfonate at high temperature. This may imply that hexadecylamine is unstable and is apt to decompose at high temperature. When coal liquefaction proceeds as the temperature reaches as high as 350°C, calcium lignosulfonate, among selected surfactants, is the best candidate for this process.

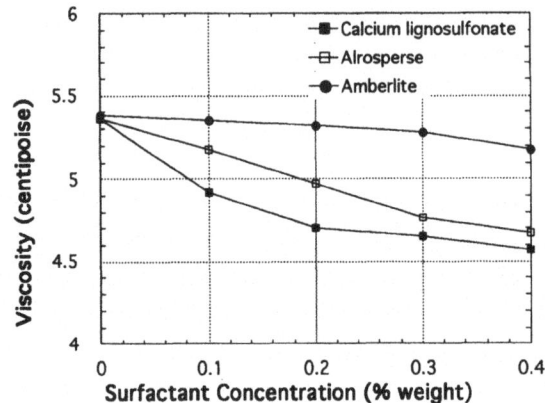

Figure 3. Viscosity of Texas Lignite coal/coal solvent
suspensions with different content of surfactants
(solid square: Calcium lignosulfonate; open
square: Alrosperse; solid circle: Amberlite).

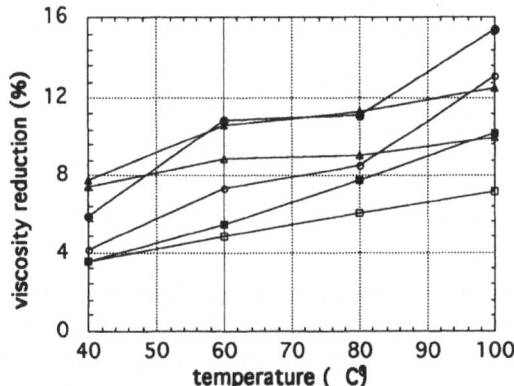

Figure 4. Viscosity reduction due to temperature and
surfactant type and content change (solid circle:
0.2% calcium lignosulfonate; open circle: 0.4%
calcium lignosulfonate; solid square: 0.2%
Alrosperse; open square: 0.4% Alrosperse; solid
triangle: 0.2% hexadecylamine; open triangle:
0.4% hexadecylamine).

Table II. Viscosity reduction due to the temperature and surfactant content change.

Surfactant	Viscosity Reduction (%) Temperature (°C)			
	40	50	80	100
0.2% Calcium lignosulfonate	4.18	7.32	8.47	13.12
0.4% Calcium lignosulfonate	5.88	10.88	11.13	15.44
0.2% Alrosperse	3.56	4.86	6.02	7.13
0.4% Alrosperse	3.58	5.41	7.77	10.21
0.2% Hexadecylamine	7.43	8.83	9.02	9.98
0.4% Hexadecylamine	7.70	10.65	11.32	12.48

AUTOCLAVE REACTION: In the surfactant screening studies, Aparently, the lignin sulfonate is the best candidate studied. Based on those findings and practical considerations, lignin sulfonate and Triton x-100 were tested in the autoclave reactor experiments. Figures 5 and 6 are the products distribution for Texas lignite and Illinois # 6 coal on a received basis. The conversion in these figures indicate the total yield of both oil and asphaltene. The results indicate a change in the distribution of light and heavy fractions. Since lignite is more refractory than coal, it is reasonable to obtain less conversion for lignite sample than the pulverized coal sample to oil and asphaltene.

For Texas lignite sample, 21% of oil and 35% of conversion are obtained after reacting at 350°C and 1900 psi hydrogen pressure for one hour. When 1% lignin sulfonate is added, the yield of oil and conversion are increased to 45% and 60%, respectively, but when 1% triton x-100 is added, the yield of oil and conversion only increase up to 34% and 49%. It is obvious that both surfactants can enhance the reaction significantly. In addition, lignin sulfonate is better for lignite system than triton x-100.

The similar results are obtained for bituminous coal sample. With the aid of 1% of lignin sulfonate, the yield of oil increases from 27% to 53% and 47% to 73% for conversion. If 0.5% lignin sulfonate and 0.5% triton x-100 are added, both oil yield and conversion of reaction would be 15% less than when lignin sulfonate is used. In this study, recycled surfactant also has been examined. The results show that recycled surfactant still can enhance the reaction

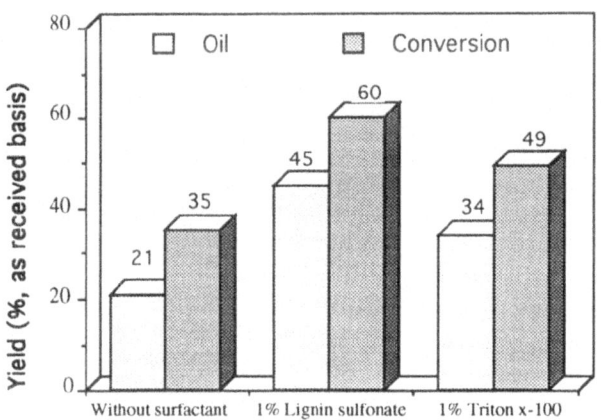

Figure 5. The autoclave products distribution for Texas lignite as received basis.

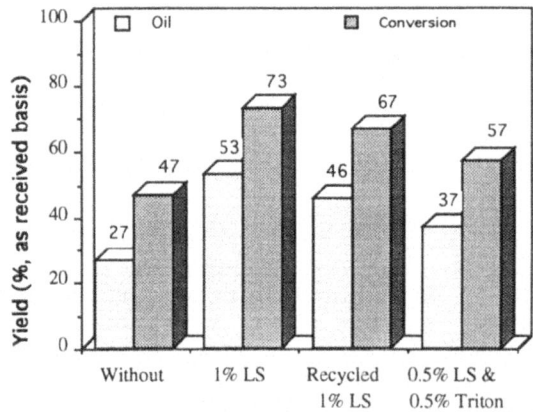

Figure 6. The autoclave products distribution for Illinois #6 bituminous coal as received basis.

significantly. It may illustrate that lignin sulfonate is very stable at severe conditions and only a small portion of lignin sulfonate is decomposed during the reaction. It definitely benefits the reduction operation cost.

Since coal contains moisture and inorganic compounds, it is impossible to convert coal completely. Figure 7 represents the results after autoclave reactions on a moisture-ash free (MAF) basis. On the MAF basis, 71% and 75% of organic components in the bituminous coal and lignite samples are converted into oil fraction when 1% of lignin sulfonate is added. Since the conversion for both coal samples reaches almost 100%, both samples can be converted within one hour.

With the participation of surfactant, the viscosity reduction and the suspended concentration increase may indicate the reduction of coal particle agglomeration. In this circumstance, the hydrogen is much easier to access, thus proceeding to coal liquefaction. This assumption has been demonstrated by the autoclave reactor experiment. Figure 8 represents the ratio of oil yield of system with surfactant to one without surfactant. Under the same condition, almost twice the yield is obtained for both samples by adding surfactants. In addition, it shows that a higher oil yield ratio is obtained for lignite sample than for bituminous coal sample. It may indicate that the addition of surfactant will affect the selection of the path of coal liquefaction.

CONCLUSION

In surfactant screening test, viscosity and suspended coal concentration for two coal samples have been conducted to roughly determine the performance of selected surfactants. All selected surfactants reduce the viscosity of system and increase the suspended coal concentrations. This may indicate that surfactants reduce the agglomeration of coal particles. The sequence of the performance of selected surfactants for two coal samples are similar. This may imply that the surface characters are similar between two coal samples.

From the temperature effect studies, it is apparent that the surfactant functions better at a higher temperature. This may be due to the fact that the polymeric surfactant will cause more bridging at low temperature than high temperature. In addition, the performance of surfactant affected by the temperature has been demonstrated. Thus, the selection of surfactant will depend on its reaction environment. At high temperature, calcium lignosulfonate is predicted to be the best among selected surfactants.

The autoclave experiments demonstrate that the addition of surfactants does not only enhance the coal liquefaction but also change the distribution of products. Surfactant lignin sulfonate has a better performance than triton x-100 on both coal samples.

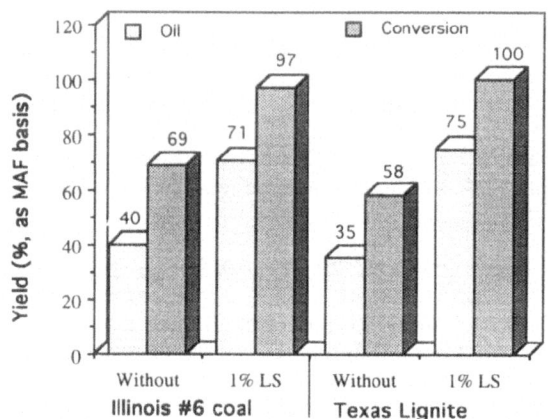

Figure 7. The autoclave products distributions for Illinois
6 coal and Texas lignite samples as MAF basis.

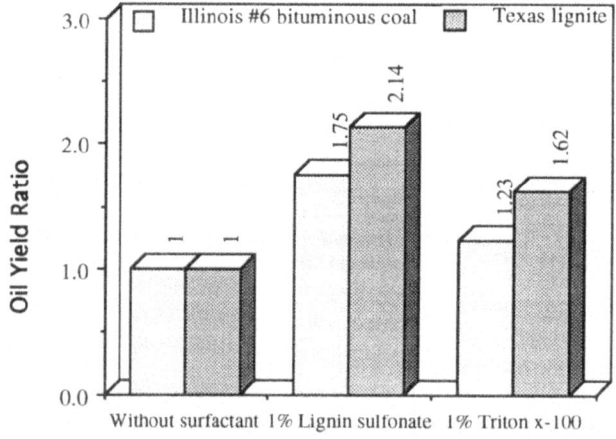

Figure 8. The enhanced oil yield by surfactants for autoclave
experiments at 350°C, 1900 psi and retention time
for 1 hour.

ACKNOWLEDGEMENT

   PARTIAL FUNDING OF THIS RESEARCH IS FROM THE JET
PROPULSION LABORATORY, CALIFORNIA INSTITUTE OF TECHNOLOGY,
PASADENA, CA.

REFERENCES

1. Characterization of Pyrolysis and on Coal Pyrolysis
    Modeling, Preprints, American Chemical Society, Division
    of Fuel Chemistry, **34(4)**, 1054-1370 (1989).

2. Berkowitz, N.; "An Introduction to Coal Technology,"
    Academic Press, New York (1980).

3. Cooper, B.R. and Ellingson, W.A. (editors); "The Science
    and Technology of Coal and Coal Utilization," Plenum
    Press, New York (1984).

4. Whitehurst, D.D. Mitchell, T.O. Farcasiu, M.; "Coal
    Liquefaction," Academic Press, New York (1980).

5. Weller, S. Pelipetz, M.G. and Friedman, S.; Kinetics of
    Coal Hydrogenation: Conversion of Asphalt. Industrial
    and Engineering Chemistry, **43(7)**, 1572-1575 (1951).

6. Shalabi, M.A., Baldwin, R.M., Bain, R.L., Gary, J.H. and
    Golden, J.O.; Noncatalytical Coal Liquefaction in a Donor
    Solvent. Rate of Formation of Oil, Asphaltenes, and
    Preasphaltenes, Industrial and Engineering Chemistry:
    Process Design and Development, **18(3)**, 474-479 (1979).

7. Yen, T.F.; Structural Difference Between Asphaltenes
    Isolated from Petroleum and from Coal Liquids, in:
    "Chemistry of Asphaltene edited by Bunger, J.W. and
    Li, N.C.)," American Chemical Society, Washington D.C.,
    pp39-51 (1981).

8. Rosen, M. J.; "Surfactants and Interfacial Phenomena,"
    John Wiley & Sons, New York (1989).

9. Elworthy, P. H., Florence, A. T. and Macfarlane, C. B.;
    "Solubilization by Surface-Active Agents. " Chapman &
    Hall Ltd, London (1968).

10. Shah, D. O. and Schechter, R. S.; "Improved Oil Recovery
    by Surfactant and Polymer Flooding," Academic Press, New
    York (1977).

11. Huggins, M. L.; The Viscosity of Dilute Solutions of
    Long-Chain Molecules, Journal of Physical Chemistry,
    42, 911-919 (1938).

12. Pal, R. and Rhodes, E.; Viscosity/Concentration
    Relationships for Emulsions, Journal of Rheology,
    33(7), 1021-1045 (1989).

STRUCTURE AND INTERACTION OF ASPHALTENE

COLLOIDS IN ORGANIC SOLVENTS

Eric Y. Sheu, M. M. De Tar, and D. A. Storm

Research and Development Department
Texaco
P. O. Box 509
Beacon, New York 12508

ABSTRACT

A series of small angle scattering measurements were performed to study the structure and the interactions of the asphaltene colloids in toluene/pyridine mixtures. In the dilute concentrations regime, the data were analyzed using a self-consistent algorithm, neglecting the interparticle interactions. The results suggest that the asphaltene colloids are approximately spherical, but with an appreciable size polydispersity. For the moderate to high concentration range, the Derjaguin-Landau-Verway-Overbeek (DLVO) potential was used to describe the interparticle interactions, and found to be adequate.

INTRODUCTION

The formation of asphaltene colloids has been a controversial issue for several decades. There are two major interpretations for the high apparent molecular weight of asphaltenes measured by gel permeation chromatography (GPC) or vapor pressure osmometry (VPO). Some interpret an asphaltene to be a huge molecule consisting of a number of polynuclear aromatics; others believe that the high molecular weight is due to self-association of asphaltene molecules, and the true molecular weight is much lower. In a previous paper[1], we showed that the colloid-like asphaltene particles suspended in organic solvents were indeed formed by self-association, with a process similar to the surfactant micellization process. The same result was also concluded by other studies, using different techniques[2,3].

Since asphaltene colloids are formed by self-association, it is important to understand the structure of the colloids and their inter-correlations at finite concentrations. Among the available techniques, small angle neutron scattering (SANS) appears to be an appropriate technique for this purpose. Several structural studies for asphaltene colloids have been published[4,5], however no conclusive evidence for asphaltene structure has been provided to date. This is partially due to the complicated nature of asphaltene colloids, and partially due to the lack of a self-consistent analysis scheme for treating SANS data. In addition to the structural study, the interparticle correlations should also be considered in the data analysis process, especially for moderate and high concentration regime. The interparticle correlation was never taken into account in the previous asphaltene study[4,5]. This is because the asphaltene colloids are expected to have diffused surfaces and there has been no appropriate potentials to describe the interactions.

In this study, an attempt was made to address both structure and interaction of this complicated system simultaneously. Since the analysis involved structure and interaction, the data fitting may be nonunique, due to too many free parameters. To avoid this to happen, we used an indirect constraint. This constrain required the particle-solvent contrast to remain constant along the concentration axis, provided the particle size distribution presumed is appropriate. In this way, the analysis can be justified through the comparison of the extracted contrast as a function of asphaltene concentration. This self-consistent algorithm is crucial for extracting the structural information from the SANS data. As for the interparticle interactions, we chose the Derjaguin-Landau-Verway-Overbeek (DLVO) potential, which consists of a spherical hard core and a Yukawa diffused potential. This is because the structure factor, a measurable quantity in SANS, can be obtained analytically by solving the Ornstein-Zernike (OZ) equation using a mean spherical approximation (MSA) ansatz[6].

Our results show that the asphaltene colloids are more or less spherical, but with considerable polydispersity, which would not be found if the interparticle interactions is not taken into account[7,8]. The DLVO potential used in our analysis seems to be appropriate.

EXPERIMENTAL

Sample Preparation

Asphaltenes were derived from a Ratawi Vacuum Residue (VR) (Neutral Zone), by a solvent fractionation procedure. The VR was dissolved in heptane in a ratio of 1 gram of VR to 40 cc of heptane (HPLC grade), the solution was stirred overnight and was subsequently filtered using Whatman no. 5 filter paper. The insoluble portion (or asphaltene fraction) was dried under a stream of nitrogen to remove residual heptane. Mass balances were performed to ensure complete solvent removal.

In order to study the asphaltene colloidal structure, as a function of solvent polarity (in terms of solvent permittivity ϵ), we dissolved asphaltenes in a series of toluene (ϵ = 2.4) and pyridine (ϵ = 12.2) mixed solvents of various volume ratios. These solutions were aged at room temperature for several days to ensure thermodynamic equilibrium.

SANS Experiment

The SANS measurements were conducted on the small angle diffractometer (SAD) at Argonne National Laboratory. The spectrometer was adjusted to cover 0.008 to 0.3 Å^{-1} scattering vector range. The sample cell for SANS measurement was a quartz cell of 1 mm path length. The temperature was maintained at 25 °C for all measurements. The absolute intensity was obtained (the differential cross-section per unit sample volume), using water as a standard and a standard data reduction procedure, provided by Argonne National Laboratory.

SANS Intensity

The intensity detected at an angle θ, in a SANS measurement represents the differential cross-section per unit volume of the sample, subtended to the solid angle that corresponds to θ. For a dispersed system, this differential cross-section is contributed from two interactions between the incident neutrons and the sample. The first interaction is due to the interference of the incoming neutron wave with the nuclei of the suspended particles. The other interaction comes from the interference between the neutron wave and the fluctuation due to interparticle interactions. Because the suspended particles are well dispersed the intensity can be written in the following form[9]:

$$I(Q) = N_p (\Delta\rho)^2 V_p^2 P(Q) S(Q) \qquad (1)$$

where N_p is the number density of the particle, $\Delta\rho$ is the scattering contrast between the particles and the solvent, V_p is the particle volume, $P(Q)$ is the intraparticle structure factor (also known as the form factor), and $S(Q)$ is the interparticle structure factor. $P(Q)$ in Eq.(1), results from the interactions between the incident neutrons and the suspended particles, it therefore carries the particle size and shape information. On the other hand, $S(Q)$

results from the interactions between suspended particles, and carries information about the potential and the strength of the interparticle interactions.

In view of Eq. (1), it becomes clear that in order to extract structural and interaction parameters from a SANS spectrum, it is necessary to model $P(Q)$ and $S(Q)$ appropriately. For a typical geometry, such as a sphere, cylinder, or ellipsoid, etc., the analytical forms for $P(Q)$ are known[9]. $S(Q)$ however, can be in different forms depending upon the physical conditions of the system under investigation. For example, an Ornstein-Zernike form is usually employed for $S(Q)$ when studying the critical phenomenon, but the Percus-Yevick $S(Q)$ form is used for a hard sphere system.

For the asphaltene colloids we are dealing with, two problems arise: (1) asphaltene colloids are highly polydispersed, and (2) asphaltene colloids have rough surfaces[7], which may potentially result in interpenetration of the asphaltene colloids, equivalent to a short range attractive interaction potential. Because of these complications, Eq. (1) needs to be modified to take into account the effect of polydispersity, additionally an appropriate interparticle potential is necessary, in order to develop the proper $S(Q)$ form.

In order to analyze the SANS spectra for asphaltene colloids, we rewrote Eq. (1) as follows:

$$I(Q) = N_p (\Delta \rho)^2 V_p^2 \langle P(Q) \rangle \langle S(Q) \rangle \qquad (2)$$

where:

$$\langle P(Q) \rangle = \int f(R) P(Q,R) \, dR \qquad (3)$$

and

$$P(Q,R) = \int g(\mu) P(Q,R,\mu) \, d\mu \qquad (4)$$

$P(Q,R,\mu)$ is the orientation averaged form factor, $g(\mu)$ is the orientation probability function, $f(R)$ is the polydispersity distribution function and R is the geometric axis where the polydispersity occurs. As one can see, one way of obtaining polydispersity and interaction is to directly invert the $I(Q)$. However, no analytical form is available for inverting the scattering kernel. A numerical inversion is necessary, if the particle size distribution is to be determined without involving presumed model. This may result in slow converging process, but also may introduce inaccuracy, due to finite data points[10]. In this study, we used an indirect method. This is by presuming a particle size distribution, compute the scattering intensity using an adjustable parameter to represent the pre-factors (number density, contrast and particle volume). Then, compute the average contrast using the extracted pre-factor value, according to Eq.(2). Since the contrast is expected to be independent of concentration, it is reasonable to argue that the particle structure and their size distribution are appropriate, if the derived contrast is indeed independent of concentration, otherwise, a new set of particle structure and size distribution is needed.

As for calculation of $<S(Q)>$, it is difficult, in the present time, to incorporate the effect of polydispersity. In fact the only available $<S(Q)>$ in an analytical form is for spherical particle with Schultz distribution in size, and with hard sphere potential[11]. Fortunately, the effect of polydispersity on $S(Q)$ is generally negligible[12]. We therefore computed $S(Q)$ using a monodisperse model with size taken as the average particle size. The potential we used was a DLVO potential which is a more flexible potential than Coulomb or exponential potential. To obtain $S(Q)$ using this potential, one needs to solve the Ornstein-Zernike equation. This has been done analytically within mean spherical approximation (MSA) ansatz, which makes the analysis process simpler. The asphaltene colloidal systems we dealt with can be consider as an one component system suspended in a continuous medium described by the solvent permittivity. The Ornstein-Zernike equation for such a primitive system can be written as

$$h(x) = c(x) + (\frac{6}{\pi}) \phi \int_V c(x') \ h(|x-x'|) d^3x' \qquad (5)$$

where h(x) is the total correlation, taking the correlations from all the particle in the solution, c(x) is the direct correlation, describing the correlation between the two referenced particles, $x = r/\sigma$ (σ = particle diameter) is a convenient dimension less reduced parameter, and ϕ is the volume fraction of the asphaltene colloids in the solution. In order to solve Eq.(5) and obtain the total correlation function, one needs a closure. The closure we used was to based on the MSA ansatz, which approximates the direct correlation function, for $X > 1$, by a potential. The potential selected here was the DLVO potential, which has been demonstrated to be an appropriate potential for many micellar and microemulsion systems[13,14]. The DLVO potential can be expressed by

$$V_{DLVO} = \gamma_o \ \frac{e^{-kx}}{x} \qquad (6)$$

where $\gamma_o e^{-k}$ is the contact potential. The structure factor, $S(Q)$, is related to $C(Q)$, the Fourier transform of c(r), by

$$S(Q) = \frac{1}{1-(6/\pi)\phi C(Q)} \qquad (7)$$

Thus, the task is to solve OZ equation and obtain c(x) for x < 1, and then Fourier transform c(x) to C(Q) for calculation of S(Q). The derivation of S(Q) for DLVO potential was achieved, and an simple analytical form was obtained[15].

RESULTS AND DATA ANALYSIS

We systematically analyzed the SANS data, according to the methods described in the previous section for <P(Q)> and S(Q) calculations. Many structural and polydispersity models were tested, and found that a spherical structure with a particle size distribution according to the Schultz distribution (see appendix A for details of this distribution) was an reasonable model for the asphaltene colloid systems we dealt with. The DLVO potential used was also appropriate. Figure 1 shows the scattering spectrum, together with a analysis (solid line). The fitting quality was reasonable. The extracted structural and polydispersity parameter are recorded in Table I.

TABLE I. EXTRACTED AND POLYDISPERSITY PARAMETERS FROM SANS FITTING

conc. (wt%)	Toluene/Pyridene*	<R> (Å)	Polydispersity (%)	K Constant
8	100/0	34.6	23.8	28.8
8	85/15	35.5	30.9	19.3
8	70/30	34.8	45.7	12.0
8	55/45	34.2	51.5	8.8
8	30/70	33.0	53.9	7.7
4	80/20	30.8	18.6	---
4	60/40	31.6	19.7	---
4	40/60	31.1	20.4	---
4	20/80	30.7	21.3	---
2	80/20	31.6	19.0	---
2	60/40	31.8	19.4	---
2	40/60	31.3	19.6	---
2	20/80	30.8	20.2	---
1	80/20	31.6	19.7	---
1	60/40	31.5	19.0	---
1	40/60	32.0	19.7	---
1	20/80	31.0	19.0	---

*Volume ratio of toluene to pyridine

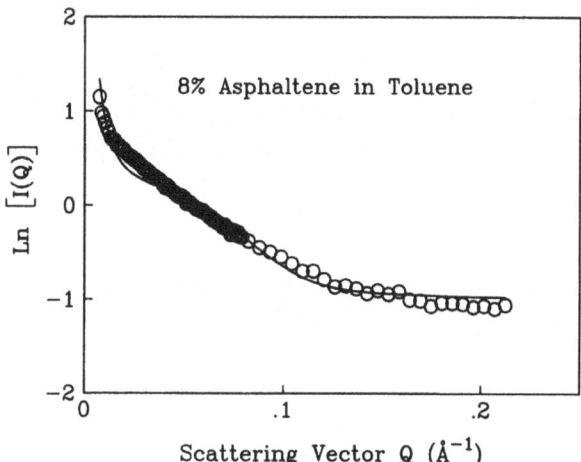

Figure 1. SANS spectrum for 8% asphaltene in toluene. The
open circles are the experimental data and the
solid line is the fitted curve, taking asphaltene
colloids as spheres, Schultz distribution as the
particle size distribution, and DLVO potential as
the interparticle correlation function.

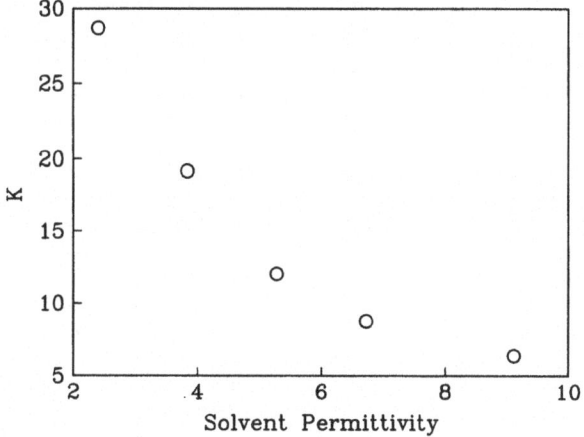

Figure 2. The decay constant k of the DLVO potential (see
Eq.(6)) as a function of the solvent
permittivity. The monotonically decreasing k
indicates that the interactions between
asphaltene colloids become longer ranged as the
solvent permittivity increases.

For asphaltene concentration less than 4%, the systems were assumed to be in the dilute regime, and $S(Q)$ was taken to be unity. The analysis for these cases involved only

$<P(Q)>$. The radius extracted did not seem to depend on either the asphaltene concentration, or the solvent permittivity (see Table I). However, strong dependence was observed for the potential diffusiveness, k (Figure 2), and particle polydispersity (Figure 3) as a function of solvent permittivity. This indicates that both particle structure and interparticle interactions are affected by solvent permittivity, this has not been observed previously by techniques such as vapor pressure osmometry (VPO), size exclusion chromatography (SEC), or even small angle scattering (up to date, the small angle scattering analysis does not take into account the interparticle interactions). The details of the data interpretation will be discussed in the following section.

DISCUSSION AND DATA INTERPRETATION

Three major features were observed in this study. First, the solvent permittivity has little impact on changing the asphaltene size. This indicates that the asphaltene self-association force is much stronger than the interaction energy between asphaltenes and solvent molecules, for solvent permittivity ranging from 2.4 to 12.2. This is expected, because a previous study[16] showed that the asphaltene colloidal size remained nearly unchanged for temperature up to 167 °C, which gave much more energy to system than adjusting the solvent permittivity between 2.4 and 12.2. Secondly, the k value was found to decrease as solvent permittivity increased. This means that the interactions between asphaltene particles become longer ranged as a function of ϵ. This observed phenomenon can be explained by assuming the solvent hydrophilicity increases as ϵ, equivalent to increasing the Debye screening length[17]. Therefore, the interaction becomes longer ranged.

In addition, the interaction was found to be repulsive, unlike that between microemulsion droplets in an oil continuous media[13], although interparticle penetration is expected. This result, however, is consistent with our previous viscosity study, using the Grimson-Barker equation[18]. Finally, the polydispersity and the size distribution (Figure 4) varies substantially as ϵ increases. A rational explanation for this phenomenon is that there are two interaction mechanisms between asphaltenes, on is strong, which makes the basic associated "unit" with a radius nearly independent of ϵ, like we observed in Figure 2. The other results in the formation of asphaltene clusters made by the associated units through a weaker force, which can be destroyed by either increasing temperature[16,19] or decreasing solvent permittivity (Figure 4). In fact, this force is likely the Van der Waal force. The polydispersity characteristics suggest that the solvent quality is enhanced as ϵ is reduced. This can be understood from Gibb's equilibrium argument, which predicts the polydispersity to decrease as solvent quality increases[14]. As we mentioned earlier, the

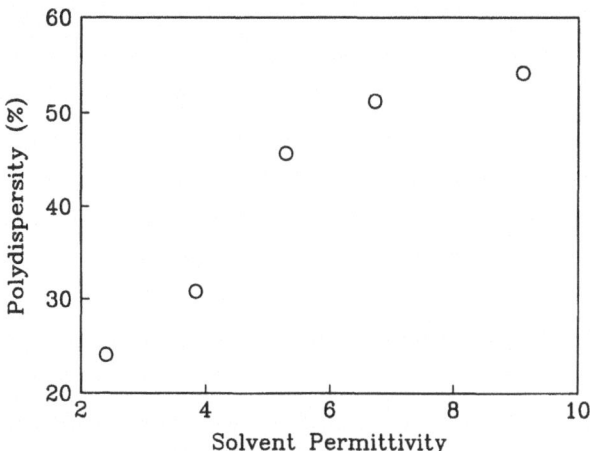

Figure 3. Particle polydispersity as a function of solvent
permittivity. It increases with increasing
solvent permittivity, indicating that toluene is
a better solvent for asphaltenes than pyridine
based on the Gibbs equilibrium argument (see text
and reference 14).

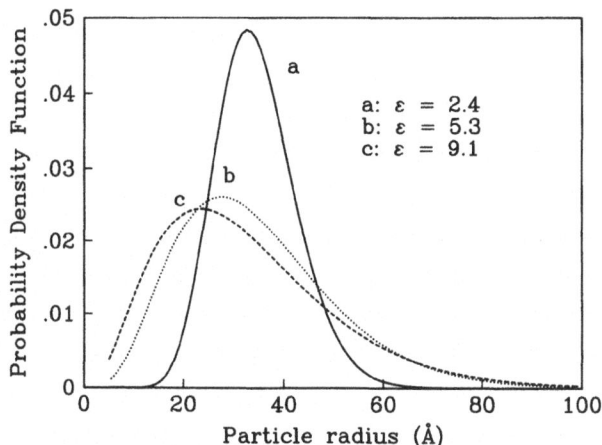

Figure 4. Size distributions for 3 ϵ values. The
polydispersity clearly increase as ϵ
increases. By lowering ϵ, one is able to
"dissolve" the clusters from by the
association unit, but the basic units remains.

evolution of polydispersity as a function of ϵ can not be observed, if the interactions between asphaltene colloids are not taken into account during SANS data analysis. It certainly can not be detected by techniques such VPO or SEC.

CONCLUSIONS

We have demonstrated that structure, polydispersity, and the interactions between asphaltene colloids in organic solvents can be obtained by proper analysis of the SANS data. The asphaltene colloid were found to be spherical, the size distribution was similar to a Schultz distribution function, and the polydispersity was greatly enhanced as ϵ increased. The interactions between asphaltene colloids can be described by the DLVO potential, and the range of interaction increases as ϵ increases. There may be two interactions involved in asphaltene self-association, one is strong, resulting in formation of self-association unit, the other is weak, resulting in clustering of the units.

ACKNOWLEDGEMENT

The author would like to thank Intense Pulsed Neutron Source (IPNS) of Argonne National Laboratory for granting small angle neutron beam time. In particular, we would like to thank D. Wozniak for helpful assistance. IPNS is operated under the auspices of US Department of Energy, BES-Material Sciences, under contract no. W-31-109-ENG-38, to whom thanks are extended for the use of the facilities.

APPENDIX

(A): Schultz Distribution Function

The Schultz distribution function is often used to describe the molecular weight distribution of a polymer system. It is a convenient function because its nth moment can be derived analytically and is a simple algebraic form. This distribution function consists of two parameters name, the average size $\langle R \rangle$ and the width parameter z. The Schultz distribution reads,

$$f(R,z) = \frac{R^z}{\Gamma(z+1)} \left[\frac{(z+1)}{\langle R \rangle} \right]^{z+1} \exp\left[-\frac{(z+1)R}{\langle R \rangle} \right] \qquad (A.1)$$

where $\Gamma(x)$ is the gamma function. The particle polydispersity can be expressed in terms of z as

$$P = \frac{\sqrt{\langle R^2 \rangle - \langle R \rangle^2}}{\langle R \rangle} = \frac{1}{\sqrt{z+1}} \qquad (A.2)$$

164

REFERENCES

1. Sheu, E. Y., De Tar, M. M., Storm, D. A. and DeCanio, S. J.; Aggregation and Kinetics of Asphaltenes in Organic Solvents, Fuel, 71, 299 (1992).

2. Anderson, S. I. and Birdi, K. S., J. Colloid Interface Sci., 142, 497 (1991).

3. Menon, V. B. and Wasan, D. T., Colloid and Surfaces, 19, 89 (1986).

4. Herzog, P., Tchoubar, D. and Espinat, D., Fuel, 67, 245 (1988).

5. Ravey, J. C., Ducouret, G. and Espinat, D., Fuel, 67, 1560 (1988).

6. Senetore, G. and Blum, L., J. Phys. Chem., 89, 2676 (1987).

7. Sheu, E. Y., De Tar, M. M. and Storm, D. A., Macromolec. Rpt. A28, 159 (1991).

8. Sheu, E. Y., Storm, D. A. and De Tar, M. M., J. Non-crystl. Solids, 131-133, 341-347 (1991).

9. Sheu, E. Y., Phys. Rev. A., 45, 2428 (1992).

10. Glatter, O., J. Appl. Cryst., 10, 415 (1977).

11. Griffith, W. L., Triolo, R. and Compere, A. L., Phys. Rev. A., 35, 2200 (1987).

12. Onsager, O., Ann. N.Y. Acad. Sci., 51, 627 (1949).

13. Samseth, J., Chen, S. H., Lister, J. D. and Huang, J. S., J. Appl. Cryst. 21, 835 (1988).

14. Sheu, E. Y. and Chen, S. H., J. Phys. Chem. 92, 4466 (1988).

15. Sheu, E. Y., Liang, K. S., Sinha, S. K. and Overfield, R. E., J. Colloid Interface Sci., 153, 399 (1992).

16. Chen, S. H., Sheu, E. Y., Kales, J. and Hoffmann, H., J. Appl. Cryst., 21, 751 (1988).

17. Sheu, E. Y., De Tar, M. M. and Storm, D. A., Fuel Sci. Tech. Int., 10(4-6), 607 (1992).

18. Sheu, E. Y., Liang, K. S.and Chiang, L. Y., J. Phys.(paris) 50, 1279 (1989).

DIFFUSION AND STRUCTURAL STUDIES OF ASPHALTENES

AND THEIR IMPLICATIONS FOR RESID UPGRADING

V. S. Ravi-Kumar, R. C. Sane[1], T. T. Tsotsis
and I. A. Webster[*]

Department of Chemical Engineering
University of Southern California
Los Angeles, California 90089

[*]UNOCAL Corporation
1201 W. 5th Street
Los Angeles, CA 90051

ABSTRACT

The diffusion and structural aspects of asphaltenes were reviewed. In order to improve the efficiency of the asphaltene hydrotreating process, a better understanding of the phenomena such as asphaltene structure, colloidal nature and mechanism of asphaltene diffusion through porous systems is desirable. The diffusive behavior showed that asphaltenes are polydisperse entities consisting of many components with a broad range of transport coefficients. Results demonstrated that the activation energies are typical of association/dissociation energies for individual asphaltene sheets rather than diffusion activation energies. A Schultz type distribution of the asphaltene molecular weight and a scaling relationship between diffusivity and molecular weight were observed.

---

[1]Present Address: Jacobs Engineering Inc.,
251 Lake Avenue, Pasadena,
CA 91101

Asphaltene Particles in Fossil Fuel Exploration, Recovery, Refining, and Production Processes, Edited by M.K. Sharma and T.F. Yen, Plenum Press, New York, 1994

167

INTRODUCTION

Among the various crude-oil fractions, asphaltenes, which consist primarily of macromolecular species, create serious difficulties in heavy oil upgrading. This is primarily due to their high heteroatom content and their macromolecular and colloidal nature, which result in reduced yields during their hydrotreating. To increase the efficiency of the asphaltene hydrotreating process it would be helpful to have a better understanding of the asphaltene structure and colloidal nature and the mechanism of asphaltene diffusion through porous systems. Recent publications are devoted for understanding the composition, molecular structure and colloidal behavior of asphaltenes.[1-3] Asphaltenes are reported to be dynamic colloidal mixtures of micelles. Micelles result from the association of smaller units called clusters, which are condensed aromatic sheets. The association of these smaller units depends on the conditions such as temperature, concentration, pressure etc. of the surrounding environment.

The dynamic polydisperse nature of asphaltenes implies one cannot define a unique time independent effective diffusivity. Several structural aspects of asphaltenes and their implications for resid upgrading are described in this paper. The colloidal nature and dynamic behavior of the asphaltenes are reviewed in order to better understand the mechanism of the asphaltene formation.

RESULTS AND DISCUSSION

Experiments were carried out in a diffusion cell consisting of two half-cells separated by the microporous membrane. More detailed results can be found in another publication[4]. The effective diffusivities were measured versus time through membranes of varying pore diameters. The diffusive behavior observed, results from the fact that asphaltenes are polydisperse entities consisting of many components with a broad range of transport coefficients. The dynamic nature of asphaltenes manifests itself in the effects of temperature and concentration on their transport. The experimentally measured activation energies are more typical of association/dissociation energies for individual asphaltene sheets rather than diffusion activation energies.

The way the Ni to V ratio changes with time in the LCS (Low Concentration Side) or HCS (High Concentration Side) during diffusion, and in the asphaltene species during reaction is an indicator of the relative abundance of these metals in the asphaltene components and provides insight into the aspects of the asphaltene structure that are of importance during asphaltene upgrading. The picture of the structure of asphaltenes emerging from the transport and reaction studies is consistent with the structural information resulting from their analytical investigations.

The dynamic structure of asphaltenes has significant implications for resid upgrading. Most of the metals found in asphaltenes would not be expected to enter the catalyst's porous structure if asphaltenes were not dynamic entities. That the LCS asphaltene has almost similar characteristics to the HCS asphaltene is indicative of the "dynamic nature" of the asphaltene molecule and its ability to generate itself from its fragments.

From the structural investigations of asphaltenes, Klein and coworkers[5] have developed a conceptual model of the asphaltene's structure. We used this structural model, and Monte Carlo and hydrodynamic methods of transport in order to simulate the diffusion of asphaltenes from the high concentration bulk phase into the cylindrical pore to the low concentration bulk phase.

We generated asphaltene molecules using the distributions for the structural elements like the number of unit sheets in a molecule, the number of aromatic and saturated rings, and the length of alkyl chains. We generated 50,000 such molecules and hence the molecular weight and size distributions. The transport coefficients of each individual asphaltene molecule in the pore were calculated using random walk techniques by approximating asphaltenes as oval cylinders and the pores as circular cylinders. A Schultz type distribution of the asphaltene molecular weight and a scaling relationship between diffusivity and molecular weight were observed.

The dynamic chemical interactions of asphaltenes within the bulk phase, and the hydrodynamic interactions of asphaltene molecules within the pores need to be taken into account to quantitatively determine the effective diffusivity of asphaltenes and to predict the temporal variation of their molecular weight distribution.

ACKNOWLEDGEMENT

The support of U. S. Department of Energy (US-DOE) is gratefully acknowledged.

REFERENCES

1. Dickie, J. P. and Yen, T. F.; Macrostructure of the asphaltic fractions by various instrumental methods, Analyt. Chem., **39**, 1847 (1967).

2. Yen, T. F.; "Structural differences between asphaltenes isolated from petroleum and coal liquids", In: Chemistry of Asphaltenes, Am. Chem. Soc. Adv. Chem. Series, **195**, 39 (1981).

3. Speight, J. G.; The Chemistry and Technology of Petroleum, Marcel Dekker, New York, (1980).

4. Sane, R. C., Tsotsis, T. T., Webster, I. A. and Ravi-Kumar, V. S.; Studies of asphaltene diffusion and structure and their implications for resid upgrading, Chem. Eng. Sci., **47**, 2683 (1992).

5. Savage, P. E. and Klein, M. T.; Asphaltene reaction pathways-V. Chemical and mathematical modeling, Chem. Eng. Sci., **44**, 393 (1989).

THE STUDY OF MOLECULAR ATTRACTIONS IN THE
ASPHALT SYSTEM BY SOLUBILITY PARAMETER

Jiunn-Ren Lin* and Teh Fu Yen

Civil and Environmental Engineering
University of Southern California
Los Angeles, CA 90089-2531 U.S.A.

ABSTRACT

Solubility parameter is an indicator which
is based on the energy needed to overcome all
the forces holding the molecules together.
Thus, the intermolecular forces can be
estimated by the solubility parameter. Both
one-component and three-component solubility
parameters are used to study the molecular
attraction within the asphalt system. In one-
component solubility parameter study, three
colloidal types of asphalts and corresponding
asphaltenes are compared at room temperature.
Comparing the spectra between asphalt and
asphaltene, it appears that asphaltene is the
controlling factor in the solubility of asphalt
system. From the spectra of three asphaltenes,
it may be implied that the attraction forces in
three different colloidal types of asphalt are
similar. In three-component solubility
parameter spectra, the distribution of
molecular attractions attributing from three
categories, dispersion, hydrogen bonding, and
polar interactions, are studied. In observing
the spectra, dispersion forces dominated in the
system is demonstrated to override either
polarity or hydrogen bonding.

Present Address: Resource Utilization Laboratory, Industrial
Technology Research Institute, Chutung 310, Hsinchu, Tiawan

INTRODUCTION

Asphalt is primarily obtained from the residue of petroleum refining. Due to its uniqueproperties, asphalt products have a myriad of uses.[1] It has been used primarily in paving and roofing applications.[2,3] Even though asphalt has been used in paving for quite some time, pavement failure still remains to cause a lot of problems. Consequently, a great deal of research has been reported to improve the performance of asphalt cement in the past several decades. So far, the oxidation and aggregation of asphalt systems is commonly recognized to be the main reason for pavement failure.[4,5] Due to the aggregation of asphaltenes, asphalt system loses parts of its adhesive properties.

Depending on its solubility in certain solvents, asphalt is generally fractionated into four important fractions: saturates, aromatics, resins, and asphaltenes (by either the SARA method or the ASTM D4124 procedure).[6,7] Saturates and aromatics together are generally considered as gas oil. Asphaltene dispersed in gas oil with resins as peptizing agents forms an asphalt system. Asphaltene will form micelles and aggregate into either super giant micelles or liquid crystals due to the composition of asphalt system.[8] The polarity of asphalt systems is conventionally perceived as the dominant factor in molecular interaction because of the presence of certain polar functional groups (S, N, and O).

Asphalt is a composite material. Asphalt has to modify its physical properties by mixing with other materials before applied in the field. Since asphalt pavements contain fillers and additives as well, the incompatibility of these materials will produce the mechanically weak points which the cracking of asphalt pavement started. In addition, the incompatibility will enhance the aggregation of asphaltenes. Thus, the compatibility of asphalt with additives is not negligible.

Generally speaking, the process of solubility is dependent on the forces of attraction and repulsion within and around the molecules in a solution. Those forces in the solvent attract the solute molecules, which at first slowly surround before entering the lattice structure. The process speeds up until a new solvent/solute complex lattice structure is formed and all the forces of attraction and repulsion reach a state of equilibrium. The lattice structure will vary according to the nature of solvent and solute, concentration, and temperature.

A satisfactory method for estimating the power of a solvent is the solubility parameter, which is based on the energy needed to overcome all the forces holding the molecules together. According to statistical mechanics, an asphalt sample can be well-dispersed if the solvent has the same or nearly the same solubility parameter. On the other hand, when the solvent has a vastly different solubility parameter, it will only be dispersed in a small extent.

Aggregation of asphalt is currently recognized to be the major factor in paving failure because of oxidation. To ensure that additives in an asphalt system disperse well and function, the solubility parameter of the additives should be similar to that of the asphalt system.

The solubility parameter concept comes from the assumption that there is a correlation between cohesive energy density and mutual miscibility.[9,10] Cohesive energy density (C) is defined as molecular cohesive energy (-E) per unit volume (V).

$$C = -E \, / \, V \qquad\qquad (1)$$

Compared to the vapor phase molecule, there are strong attractive forces between molecules and considerable potential energy in the condensed phases (solids, liquids, or solutions). Molecular cohesive energy is the energy required for a complete isothermal vaporization of a saturated liquid to gas at zero pressure (i.e., infinite separation of the molecules or ideal conditions). V is the molar volume of the liquid. The solubility parameter (δ) is defined as the positive square root of the molecular cohesive energy density:[12,13]

$$\delta = C^{1/2} = (-E \, / \, V)^{1/2} \qquad\qquad (2)$$

From thermodynamic point of view, the Gibbs free energy of mixing of the system, ΔG_m, can be expressed as equation 3. While a material is dissolving in a solvent, the Gibbs free energy of the process must be negative.

$$\Delta G_m = \Delta H_m - T\Delta S_m \qquad\qquad (3)$$

Since the entropy change (ΔS_m) is always positive, the heat of mixing (ΔH_m) determines whether dissolution will occur or not. Only when the heat of mixing is small enough, dissolution will take place. Although there is no exact formula for the molar heat of mixing, the Hildebrand-Scatchard formula represents the total heat of mixing fairly well. According to the Hildebrand-Scatchard theory,[13] the heat of mixing is given by

$$\Delta H_m = V_m \, (\delta_1 - \delta_2)^2 \, \phi_1 \, \phi_2 \qquad\qquad (4)$$

where Vm is the molecular volume of the mixture, and ϕ_1, ϕ_2 are the volume fraction of the solvent and solute respectively. Therefore, when δ_1 is close to δ_2 the heat of mixing will be either small or zero and the free energy change will be negative.

The above relation (equation 4) holds for non-polar liquids. It has been assumed that the same principle holds for polar liquids and solids, as it has proven to be true in practice.[14,16]

The cohesive energy (-E) arises from several forces. Hansen and coworkers classified cohesive energy into three categories, on the basis of energy contribution: (1) $-E_d$ for nonpolar or dispersion interactions, (2) $-E_p$ for polar interactions, and (3) $-E_h$ for hydrogen bonding including donor-acceptor or similar specific interactions.[12,14] At the same time, the solubility parameter can be expressed as three different terms, according to these contributions.

$$-E = (-E_d) + (-E_p) + (-E_h)$$

or

$$\delta_o{}^2 = \delta_d{}^2 + \delta_p{}^2 + \delta_h{}^2 \tag{5}$$

Sometimes, the last two terms of equation 5 are collectively described as the "polar solubility parameter," δ_a.

$$\delta_a = (\delta_p{}^2 + \delta_h{}^2)^{1/2} \tag{6}$$

EXPERIMENTAL

In the one-component solubility parameter test, the solubilities of both asphalt and asphaltene samples were determined at room temperature with different solvents. For each run, a 0.5 g of asphalt or asphaltene sample was mixed with 10 ml of solvent in a flask by a magnetic stirrer for 10 minutes. The filtration process followed, removing insoluble materials from the dispersion. The filtrate was then placed in a vacuum oven and weighed after drying. The dispersion percentage (or solubility) of the asphalt system could then be determined.

As for the three-component solubility test, a sample is immiscible when its solubility is less than 50 g/L, so the insoluble portion actually includes the samples that are partially soluble. To determine whether the sample was soluble in the solvent, a procedure very similar to the one-component solubility parameter test was used. The insoluble portion in the solvent is determined as to whether any residue remained in the Whatman No.1 filter paper after filtration.

Three different asphalt samples from different geographic locations and refining processes were used in this research. They include West Texas asphalt generated by solvent extraction, Lloydminister asphalt by distillation and West Texas Intermediate/West Texas Sour Asphalt by air-blown refinery. These three samples had been recognized to represent three different colloidal types of asphalt,[16] where the West Texas asphalt is a sol type asphalt, the Lloydminister asphalt is a sol-gel type asphalt, and the West Texas Intermediate/West Texas Sour Asphalt belongs to the gel type. All of the asphalt samples were obtained through the material bank of Strategic Highway Research Program (SHRP), operated by the University of Texas, Austin.

For each asphalt sample, the asphaltene fraction was obtained through precipitation followed by Soxhlet extraction. Around 10 g of asphalt sample was dissolved into 10 ml of toluene and a toluene dispersion was formed. Asphaltene precipitation occured when an excess of n-pentane was added to a toluene dispersion. A toluene-pentane ratio of 1:50 (v/v) was used to ensure precipitation of the asphaltene, whereas, Soxhlet extraction was used to ensure isolation of the asphaltene sample. n-Pentane was used during the extraction as an eluent until it became clear, which took over 48 hours. In order to avoid oxidation the samples were dried in a vacuum oven overnight at room temperature.

All the solvents used were reagent grade or better with no further treatment before usage.

RESULTS AND DISCUSSION

In the study of one-component solubility parameter spectra, three colloidal types of asphalt samples and their corresponding asphaltenes were investigated and compared. The detailed distribution of molecular forces within asphaltene has illustrated in the section of three-component solubility parameter spectra.

ONE-COMPONENT SOLUBILITY PARAMETER SPECTRA

To obtain the solubility parameter spectra, some solvent mixtures were used in the study. The solubility parameters of those solvent mixtures were calculated with the following equation:[13]

$$\delta_m = \delta_1 \phi_1 + \delta_2 \phi_2 \tag{7}$$

where δ_m, δ_1, and δ_2 are the solubility of the mixture and the two solvents respectively, and ϕ_1 and ϕ_2 are the volume fractions of the correspondent solvent. Strictly speaking, when solute (asphalt or asphaltene) is mixed with the solvent or solvent mixture, the solubility parameter of whole system should include one term coming from solute. Thus, the right hand site of Equation 7 shall include another term attributing from solute. Since the volume fraction is relatively small, this term is negligible.

As a physical constant, the solubility parameter measures the force by which solvent molecules attract one another. If a liquid and an asphaltene/asphalt system have the same or nearly the same solubility parameter, the liquid is a solvent for the asphaltene/asphalt system.

Figures 1-3 represent the solubility parameter spectra
of three colloidal types of asphalts and corresponding
asphaltenes. It is obvious to observe that the solubility
parameter spectra of asphalt are different not only within
the three colloidal types of asphalt but also significantly
from asphaltene. It is apparent that the composition of the
three different types of asphalt vary significantly. The
contents of resin and asphaltene contribute much to this
result. The resin content will enhance the dispersion of
asphaltene particles and prevent their aggregation. West
Texas asphalt (sol type) have a higher resin content and a
better dispersion percentage than Lloydminister asphalt (sol-
gel type), which in turn is so with West Texas Intermediate/
West Texas Sour asphalt (gel type); therefore, a higher
entropy change is obtained from sol type to sol-gel type to
gel type.

From the spectra of asphalt and corresponding
asphaltene, it is apparent that the range of asphaltene is
narrower than that of asphalt. It indicates that the
asphaltene control the solubility of asphalt system. If the
asphaltene can be dissolved in a certain solvent, the asphalt
system would be miscible with this solvent. From the spectra
of asphaltenes, the range of the solubility parameter and the
shape of the solubility parameter spectra for the three
samples are similar . The solvents with a solubility
parameter ranging from 8.4 to 10.5 hildebrands will disperse
the asphaltene fairly well and will not have any precipitates
in the dispersion. The differences among these three samples
are in the tail portion. It implies that the cohesive
energies of three asphaltene samples from different colloidal
types of asphaltene are fairly similar. Also, the difference
in the composition and structure of asphaltene for three
distinct types are insignificant.

THREE-COMPONENT SOLUBILITY PARAMETER SPECTRA

The studies of the critical micelle concentration (CMC)
of asphaltene demonstrated that there was an approximately
linear relationship between CMC and solubility parameter.[14,17]
This relationship is only valid for the toluene-n-alkane
system. For the solvents with different polarity and hydrogen
bonding capacity, CMC cannot be correlated with the
solubility parameter. It is reasonable for this conclusion
since all the n-alkane solvent have very minor polarity and
hydrogen bonding capacity. For these solvents with different
strengths of interaction forces, the asphaltenes will have an
increase in interactions with the solvent medium and form
into micelles with various configuration. To better
understand the asphaltene system, three-component solubility
parameter spectra is essential to reveal the interactions
among the asphaltene molecules.

Since we cannot find a formula that can express the
cohesive energy contributions (i.e., $-E_d$, $-E_p$, and $-E_h$) of the
solvent mixture, forty solvents with various cohesive energy
distribution were selected as controlling points for

constructing the three-component solubility parameter of asphaltene system. The distribution of cohesive energy sources for controlling points are shown in Figures 4 and 5.

In Figures 6 to 11, solid circles and open squares are used to represent the solvents which are either soluble or insoluble with asphaltene, respectively. From Figures 6-8, a three-dimensional diagram can be constructed to illustrate the solubility parameter area of asphaltene. With a few exception, it shows that the asphaltene is miscible with solvents with following solubility parameter range:

* solubility parameter by dispersion forces:
  > 8 hildebrands

* solubility parameter by hydrogen bonding forces:
  0.5 - 4 hildebrands

* solubility parameter by dipole-dipole forces:
  0.5 - 4 hildebrands

It is obvious that the solubility parameter by dispersion forces are much larger than the others. Since the solubility parameter is defined as the square root of cohesive energy density, the dispersion forces within the asphaltene seem to be higher than one caused by polarity and hydrogen bonding.

In Figures 9-11, the individual effect of nonpolar interaction, polar interaction and hydrogen bonding interaction on asphaltene are addressed. The relationship between overall and dispersion solubility parameters indicates that the asphaltene system is in favor of the solubility parameter which is strongly caused by nonpolar interaction. All the controlling points having dispersion solubility parameter larger than 8.5 hildebrands are miscible with asphaltene without exception. Solvent with either very weak or very strong hydrogen interaction will not be good solvents for asphaltene. It may imply that the asphaltene does contain some molecular forces by hydrogen bonding. But the strength of these forces is not as significant as dispersion forces. Only the solubility parameter caused by hydrogen bonding less than 4.0 hildebrands is suitable for asphaltene. Figure 11 shows the role of polarity forces in the overall molecular forces. The data shown are scattered, implying that the polarity effect of solubility is not as significant as the other two factors.

CONCLUSION

The distribution of molecular forces within asphalt systems have been investigated by either one-component or three-component solubility parameter spectra at room temperature.

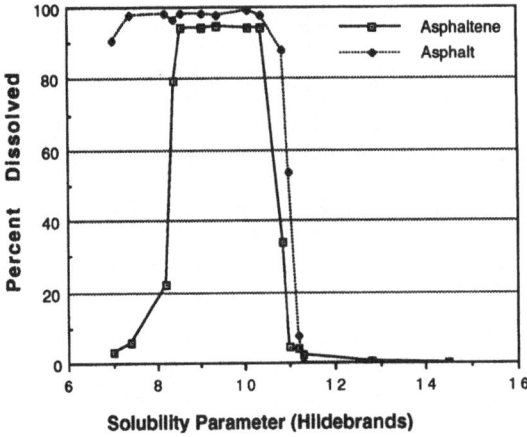

Fig. 1. The solubility parameter spectrum for West Texas asphaltene and asphalt.

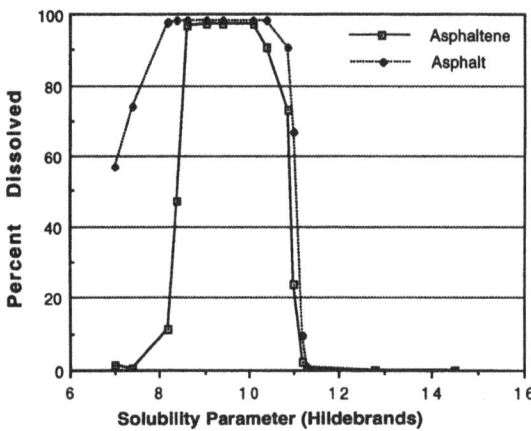

Fig. 2. The solubility parameter spectrum for Lloydminister asphaltene and asphalt.

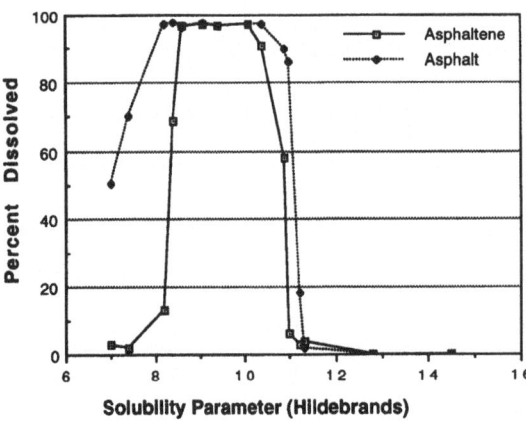

Fig. 3. The solubility parameter spectrum for West Texas Intermediate/West Texas Sour asphaltene and asphalt.

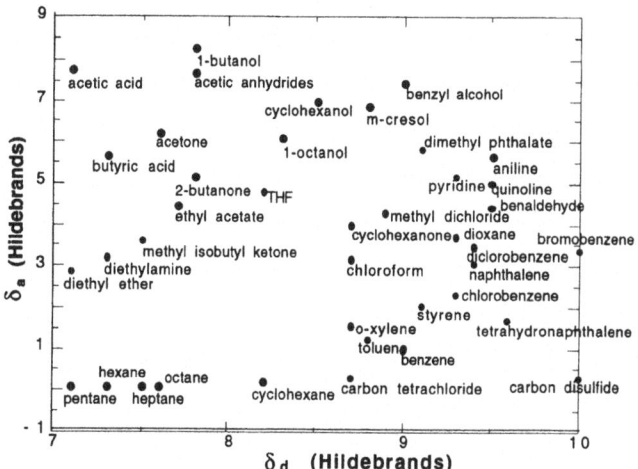

Fig. 4. The relationship between dispersion solubility parameter and polar solubility parameter of controlling compounds.

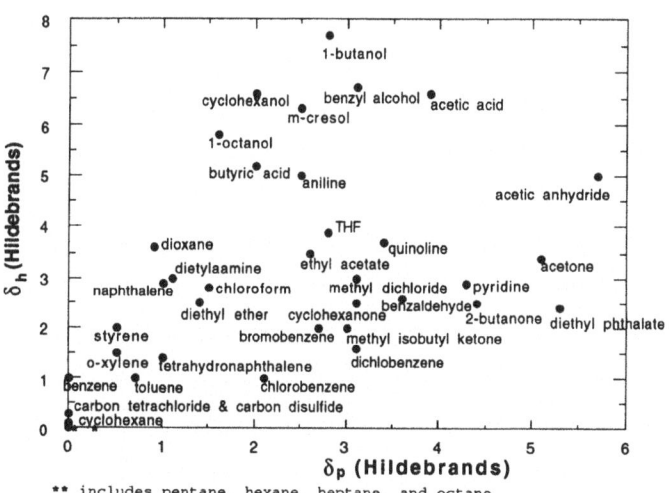

** includes pentane, hexane, heptane, and octane

Fig. 5. The controlling compounds distribution according to the solubility parameter caused by polarity and hydrogen bonding.

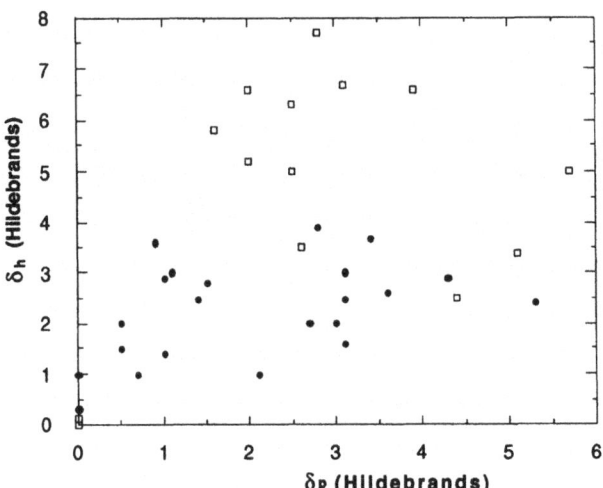

Fig. 6. The effect of hydrogen-bonding and polarity forces on the miscibility of ashpaltene sample.

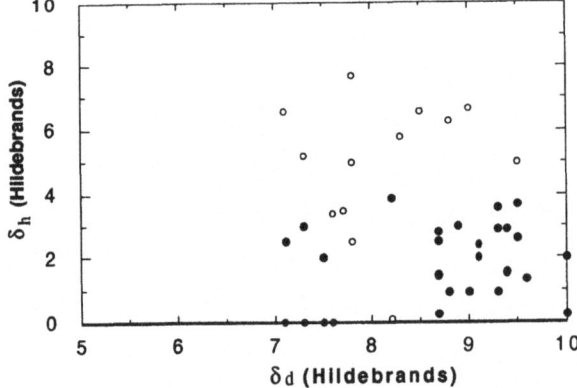

Fig. 7. The effect of dispersion and hydrogen-bonding forces on the miscibility of asphaltene sample.

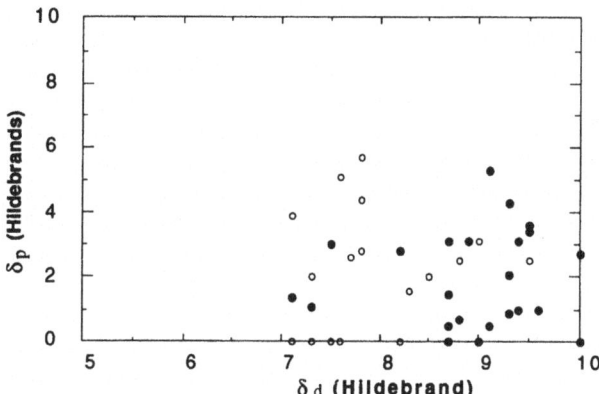

Fig. 8. The effect of dispersion and polarity forces on the miscibility of asphaltene sample.

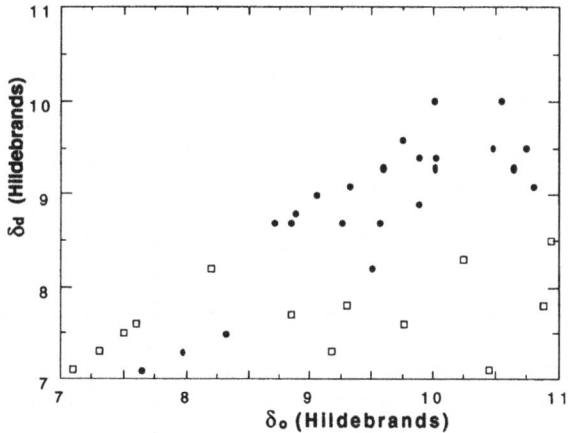

Fig. 9. The effect of nonpolar interaction on the overall solubility parameter for asphaltene sample.

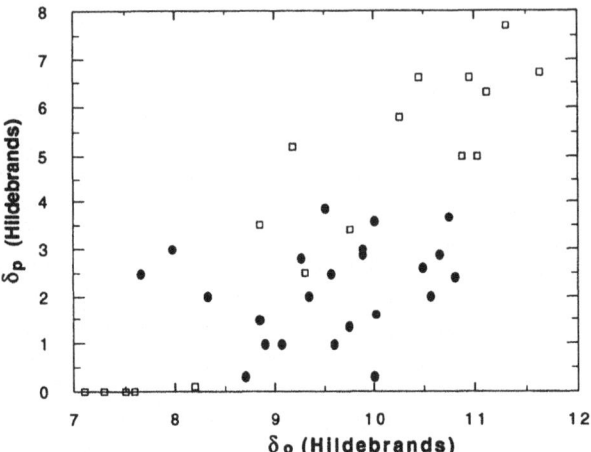

Fig. 10. The effect of hydrogen bonding interaction on the overall solubility parameter for asphaltene sample.

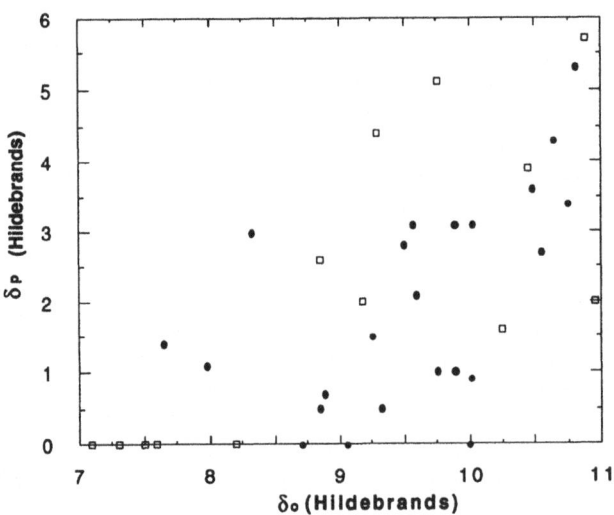

Fig. 11. The effect of polarity interaction on the overall solubility parameter for asphaltene sample.

One-component solubility parameter spectra reveal that the asphaltene have a narrower range of solubility parameter than that of asphalt. It indicates that the additive is compatible with asphalt system if it can compatible with asphaltene. In addition, three different types of asphaltene have similar solubility parameter range and shape. This may imply that the characterization and structure of these asphaltene are similar. Since the cohesive bonding comes from not only dispersion force but also from hydrogen bonding and polarity in the asphalt system, a one-component solubility parameter test is not good enough to reveal the detailed distribution of those interactions within asphalt systems.

From the three-component solubility parameter, the molecular interaction among the asphaltene can be better understood. The results indicate that the dispersion force is still the most significant intermolecular force. The cohesive energy caused by the hydrogen bonding is less than one quarter of that from the dispersion force. Compared to polarity forces, hydrogen bonding in the asphalt system is important, and it is not negligible for the study of the chemistry of asphalt. Furthermore, the data for polarity interaction reveal that this force is not as significant as other forces in controlling the behavior of asphalt systems. Therefore, the controlling interaction force of asphalt systems for its compatibility is due to dispersion and hydrogen bonding.

In order for the additives to be compatible with asphalt systems and to reduce the aggregation, solubility parameters can be used as an indicator and the additive solubility parameter range that falls into the following range is to be selected. The dispersion and hydrogen bonding solubility parameter is in the range above 8.5 hildebrands and 0.5-4.0 hildebrands, respectively.

REFERENCES

1. The Asphalt Institute, "A Brief Introduction to Asphalt and Some of Its Uses: Manual Series No. 5," 8th edition, College Park, Md. (1986).

2. The Asphalt Institute, "Report on Sales of Asphalt in the U.S. in 1981," College Park, Md. (1982).

3. Williamsburg, v.; Minutes of Committee A2D05, General Asphalt Problems, Transportation Research Board, National Research Council, Washington D. C. (1983).

4   Mill, T. and Tse, D.; Oxidation and Photooxidation of Asphalts, Preprints, American Chemical Society, Division of Petroleum Chemistry, **35(3)**:483-489 (1990).

5. Corbett, L.W.; Asphalt and Bitumen, in: "Ullmann's Encyclopedia of Industrial Chemistry," A3:178. VCH, New York (1985).

6. Altgelt, K. H., Jewell, D.M., Latham, D. R. and Selucky, M. L.; "Chromatography and Petroleum Analysis," Marcel Dekker, New York and Basel, pp194-196 (1976).

7. Standard Test Method for Separation of Asphalt into Four Fractions(ASTM D4124), in:" American Society for Testing and Materials," Philadelphia (1989).

8. Yen, T. F.; Asphaltic Materials, in: "Encyclopedia of Polymer Science and Engineering," Second edition. John Wiley & Sons, New York, pp1-10 (1990).

9. Weinberg, V. L. and Yen, T.F.; Solubility Parameters in Coal Liquefaction Products, Fuel, **59**, 287-289 (1970).

10. Barton, A. F. M.; "Handbook of Solubility Parameters and Other Cohesion Parameters," CRC Press, Baco Raton, Florida (1983).

11. Hildebrand, J. H. and Scott, R.L.; "The Solubility of Non-electrolytes," Third edition. Reinhold, New York (1950).

12. Barton, A. F. M.; Solubility Parameters, Chemical Reviews, **75(6)**, 731-753 (1975).

11. Mellan, I.; "Compatibility and Solubility," Noyes, New Jersey, (1968).

14. Hansen, C. M.; The Universality of the Solubility Parameter, Industrial and Engineering Chemistry: Product Research and Development, **8(2)**, 2-9 (1969).

15. Lin, J. R., Lian, H. J., Sadeghi, K.M. and Yen, T.F.; Asphalt Colloidal Types Differentiated by Korcak Distribution, Fuel, **70(12)**, 1439-1444 (1991).

16. Andersen, S. I. and Birdi, K. S.; Aggregation of Asphaltenes as Determined by Calorimetry, Journal of Colloid and Interface Science, **142**, 497-502 (1991).

17. Grant, D. J. W. and Higuchi, T.; "Solubility Behavior of Organic Compound," John Wiley & Sons, New York, (1990).

SOLVATION OF RATAWI ASPHALTENES IN

VACUUM RESIDUE

David A. Storm, Ronald J. Barresi and Eric Y. Sheu

Texaco Research and Development
P.O. Box 509
Beacon, New York 12508

ABSTRACT

Analyses of small angle X-ray scattering data, and the shear rheology of natural and synthetic Ratawi vacuum residue indicates that Ratawi asphaltenes(heptane insolubles) are polydispersed spheres solvated by the non-asphaltene fluid at 93 °C. The Pal-Rhodes equation attributes the non-linear dependence of the relative viscosity on asphaltene concentration to solvation and polydispersity effects. In other words the direct interactions between the asphaltene particles and the surrounding fluid, those that determine the shear viscosity, are accounted for by the solvation effect. The existence of solvation, and the amount of solvation obtained from the Pal-Rhodes analysis are consistent with analyses based on equations given by Einstein, Eilers, and Campbell and Forgacs(percolation theory). The theory of Grimson and Barker confirms that direct inter-particle interactions between solvated particles are small at 93 °C. The amount of solvation can be modified by adding other substances. Increasing the resin concentration (pentane insolubles/heptane solubles), or adding small amounts of butylamine increases the amount of solvation and viscosity. On the other hand heavy cycle gas oil(HCGO) decreases the amount of solvation. The asphaltene particles have a smaller influence on the surrounding fluid when HCGO is present, and so the shear viscosity is lower.

Asphaltene Particles in Fossil Fuel Exploration, Recovery, Refining, and Production Processes, Edited by M.K. Sharma and T.F. Yen, Plenum Press, New York, 1994

185

INTRODUCTION

When crude oil enters a refinery it passes through two distillations where fractions that are relatively easy to refine are removed. The fraction that remains after the vacuum distillation is called vacuum residue, and it is difficult to refine into gasoline, jet fuel and diesel fuel. Much of the difficulty appears to be associated with a group of molecules called asphaltenes.

Asphaltenes are a group of molecules in the vacuum residue that are defined by being insoluble in heptane, and soluble in toluene[1]. They are a complicated mixture, and both their molecular structure, and macrostructure in the vacuum residue is a subject of debate[2,3]. In this work we present evidence that the asphaltenes are spherical particles suspended in vacuum residue, about 30-50 A in radius. They are also solvated by other molecules in the non-asphaltene fraction. The size of the solvation shell can be influenced by chemical additives, such as butyl amine, heavy cycle gas oil(HCGO), or the "resins". The "resins" are defined in this work as the fraction of vacuum residue that is insoluble in pentane, but soluble in heptane. Since the shear rheology for vacuum residue is that for a simple suspension of solvated spheres, the viscosity of vacuum residue is strongly affected by additives that change the size of the solvation sphere.

EXPERIMENTAL

Samples of synthetic vacuum residue with varying amounts of asphaltenes were prepared by dispersing vacuum residue in the non-asphaltene fraction of the vacuum residue. The non-asphaltene fractions were prepared by mixing one part of vacuum residue with forty parts of heptane, and stirring overnight. The asphaltenes were then removed by filtration, and the heptane was removed from the oil phase by vacuum distillation. Samples of synthetic Ratawi vacuum residue with different amounts of resins were prepared by dispersing Ratawi vacuum residue in the fraction of the vacuum residue soluble in pentane. As with the heptane soluble fraction, pentane was removed after filtration by vacuum distillation.

The small angle X-ray scattering(SAXS) experiments were made with the ten meter spectrometer at Oak Ridge National Laboratory. The wave length of the X-rays was 1.54 A; wave vectors were in the range of 0.1-0.25 A^{-1}. The temperature of the vacuum residue samples was 93 °C in these experiments.

Viscosities were measured at several shear rates using a Couette type viscometer. All measurements were made at 93 °C. The viscosity was first measured at the lowest shear rate possible, and then at successively higher shear rates. The shear rate was then decreased until the initial shear rate was obtained. Hysteresis effects were not observed, and both the natural, and synthetic vacuum residues behaved as Newtonian fluids.

RESULTS

The vacuum residues, their asphaltene concentration, and the hetero-atom content of the asphaltenes is given in Table I.

TABLE I. COMPOSITION DATA FOR ASPHALTENES

Crude Oil	Location	% Asph.	% S	Ni (ppm)	V (ppm)
Duri	Indonesia	2.1	1.1	295	111
ANS	Alaska	6.8	3.4	209	438
Ratawi	M.E.	21.8	7.7	145	308
Oriente	Ecuador	28	4.1	375	804
Merey	Venezuela	31	4.5	336	1573

The SACS results were analyzed as previously described[3]. The asphaltenes appear to be spherical particles within the resolution of the instrument. The sizes of the particles are distributed according to a Schultz distribution[3], which in this case is very close to being Gaussian. The average radii of the asphaltenes in the various vacuum residues are given in Table II, along with the percent polydispersity. Polydispersity is defined as the square root of the difference between the average of the radius squared and the average radius squared all divided by the average radius. Basically the radii of particles were in the range of 20-80 A°.

TABLE II. SIZE AND PERCENT DISPERSITY OF ASPHALTENE COLLOIDAL PARTICLES

Vacuum Residue	Ave. Radius (A)	% Polydispersity
Duri	51.0	19.4
Ratawi	33.8	15.4
Oriente	35.1	18.2
Merey	39.2	12.6

As discussed previously, vacuum residue behaves as a suspension of solid particles[4]. It is well known that in such suspensions, the relative viscosity, which is the viscosity of the suspension divided by the viscosity of the suspending medium, depends linearly on the concentration of the particles. This results is shown in Figure 1 for Alaska North Slope vacuum residue. The slope of the line shown in Figure 1, is 2.5 times a solvation constant. The 2.5 comes about because the particles are spherical, and the intrinsic viscosity of spheres is 2.5. The solvation constant accounts for the fact that the volume of the asphaltene particles is larger in vacuum residue than in the dry state(precipitated). The concept of solvation has been shown to be consistent with several more powerful theories for the rheology of suspensions[5]. Table III shows the apparent intrinsic viscosities obtained from an analysis of the rheology in the dilute regime, and the solvation constants for these asphaltenes.

TABLE III. SOLVATION CONSTANTS FOR ASPHALTENES

V.R.	$[\eta]_W$	K(VR)	K(HCGO)	K(BA)	K(RESIN)
Duri	6.6	2.6			
ANS	6.7	2.7			
Ratawi	7.3	2.9	2.4	3.1	3.1
Oriente	8.6	3.0			
Merey	9.1	3.6			

One of the rheological theories for concentrated suspensions of solid particles is the Pal-Rhodes equation(5):

$$\eta_r = (1 - K \times W_A)^{-2.5} \qquad (1)$$

where η_r is the relative viscosity, K is the solvation constant, and W_A is the weight fraction of asphaltenes in the vacuum residue. The functional form in Eq(1) arises because the particles are polydispersed, and the exponent 2.5 arises because the particles are spherical. Figure 2 shows the Pal-Rhodes plot for the samples of natural and synthetic Ratawi

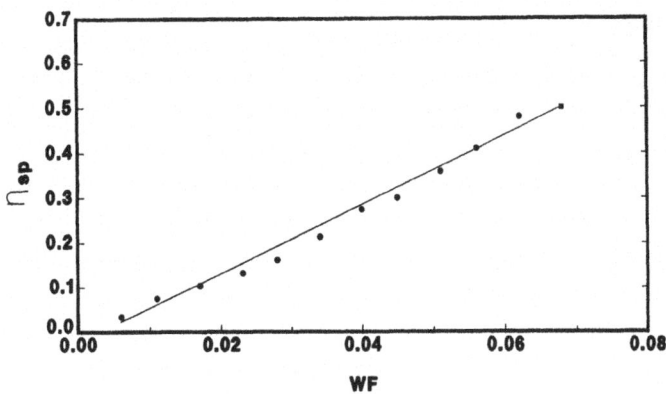

Fig. 1. Specific Viscosity (Relative Viscosity Minus One) Of ANS Versus Weight Fraction Asphaltenes.

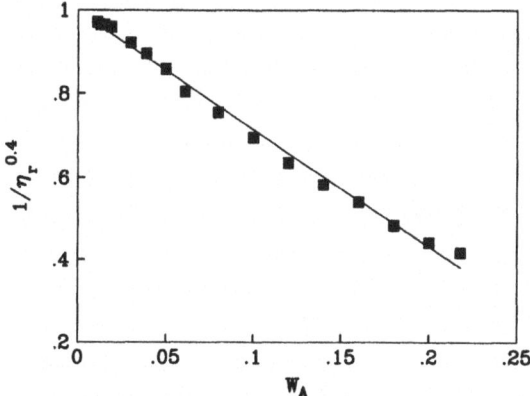

Fig. 2. Relative Viscosity For Ratawi Raised To The 0.4 Power Versus Weight Fraction Of Asphaltenes.

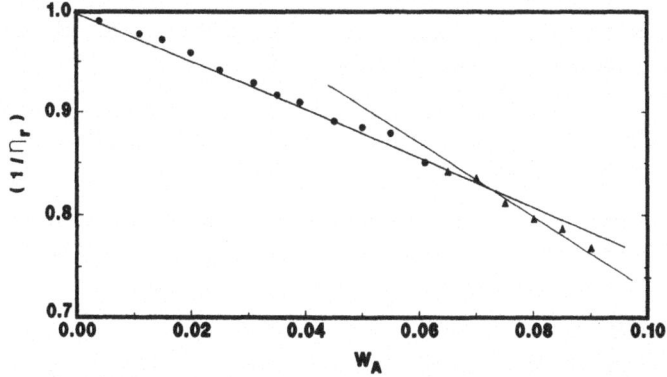

Fig. 3. Relative Viscosity For Ratawi Plus Resins Raised To The 0.4 Power Versus Weight Fraction Of Asphaltenes. Resin Concentration Varies From Essentially Zero At Low Asphaltene Concentrations To That In Vacuum Residue For An Asphaltene Concentration Of 21.8%

189

vacuum residue. Figure 3 shows the Pal-Rhodes plot when the concentration of resins in the residues varies. The dilute samples contain essentially zero resins, while the sample of vacuum residue contains approximately 7-8 % resins. The curvature in Figure 3 indicates the value of the solvation constant varies with resin concentration indicating that molecules in the resin sub-fraction of vacuum residue affect the size of the solvation sphere surrounding the asphaltene particle. The addition of HCGO and butyl amine also have a effect on the solvation; values for the solvation constants in these cases are given in Table III.

DISCUSSION

The SACS results show that asphaltenes are small spherical particles in their natural state in vacuum residue. The particles are larger than typical molecules, and in the

size range of typical micelles. It is well known that micelles form by a self-association of surfactant molecules in water; hydrophilic end groups on the outside of the micelle, while the hydrophobic tails are inside. Although the exact structure of these asphaltene colloidal particles in vacuum residue is uncertain at present, it is known that Ratawi asphaltenes self-associate to form particles in pyridene and nitrobenzene(6). The size and shape of the particles in toluene is the same as the asphaltenes in vacuum residue. The spherical shape and poly-dispersity of sizes in vacuum residue is supported by the rheological analyses above and other rheological analyses[4,5].

The rheological analyses also show the asphaltene particles are solvated by the other molecules in the vacuum residue. In principle this only means that the density of the asphaltenes in vacuum residue is less than what it is in the dry state. This could happen if the asphaltene particle swells due to the other molecules in vacuum residue, or because the other molecules in vacuum residue surround the asphaltene particle, and form a classical solvation shell. Our previous analyses indicate a classical solvation shell forms[5]. The data in Table III therefore indicates that HCGO reduces the size of the solvation shell surrounding the Ratawi asphaltene particle. The viscosity also goes down because the smaller solvated spheres can pass by each other easier in response to a change in the stress field. On the other hand, the resins and butyl amine increases the size of the solvation shell, and the viscosity of the mixture increases. Again this is understandable in terms of percolation theory; it is more difficult for the larger spheres to move into holes in the macroscopic cluster which forms in the concentrated mixture of asphaltene particles[5].

CONCLUSIONS

1. Ratawi asphaltenes are small colloidal particles suspended in the non-asphaltene portion of vacuum residue.

2. The average diameter of the asphaltene colloidal particles is approximately 100 °A.

3. The size of the Ratawi asphaltene colloidal particles is in the range for micelles, and there is some evidence that Ratawi asphaltene molecules self-associate to form these micelle-like particles.

4. The asphaltene colloidal particles are solvated by molecules in the non-asphaltene portion of Ratawi vacuum residue, and so they occupy a larger volume in vacuum residue than one would expect based on their specific volume.

5. Additives such as butyl amine, heavy cycle gas oil, and the pentane insoluble/heptane soluble fraction of the vacuum residue alter the size of the solvation shell, and hence the viscosity, since the viscosity according to percolation theory is determined by the ease at which the particles can move pass one another.

REFERENCES

1. Speight, J. C. " The Chemistry and Technology of Petroleum", Marcel Dekker, Inc., New York, p. 401 (1991).

2. Herzog, P., Tchoubar, D. and Espinat, D., "Macrostructure of Asphaltene Dispersions by Small-Angle X-Ray Scattering", Fuel, **67**, 245 (1988).

3. Storm, D.A., Sheu, E. Y. and DeTar, M. M., "Macrostructure of asphaltenes in vacuum residue by small angle X-ray scattering", Fuel (IN PRESS).

4. Storm, D. A., Baressi, R. J. and DeCanio, S. J., "Colloidal Nature of Vacuum Residue", Fuel, **70**, 779 (1991).

5. Storm, D. A. and Sheu, E. Y. "Rheological studies of Ratawi vacuum residue at 366 K", Fuel, **72**, 233 (1993).

6. Sheu, E. Y., DeTar, M. M., Storm, D. A. and DeCanio, S. J., "Aggregation and Kinetics of Asphaltenes in Organic Solvents" Fuel, **71**, 299 (1992).

AIR-BLOWN ASPHALT MADE FROM ROOM TEMPERATURE AND AMBIENT PRESSURE UNDER ULTRASOUND

H. J. Lian and T. F. Yen

Environmental and Civil Engineering
University of Southern California
Los Angeles, CA 90089-2531

ABSTRACT

In general, air-blown asphalt can be made by several methods, which employ high temperature and pressure conditions. A new method, which not only can overcome this high temperature and pressure problem, but also can effectively produce air-blown asphalt within a short time, is to increase asphaltene content rapidly by adding a surfactant into an asphalt emulsion system under ultrasound. This method can produce different types of asphalt for various usages such as sol-gel type asphalt for the encapsulation of hazardous material and gel type asphalt for metal surface coating.

INTRODUCTION

Asphalt is a very complex mixture which is obtained primarily from bottom residue of the petroleum refining process. In the United States, a stylized barrel of crude oil yields approximately 45% gasoline, 5% kerosene, 34% light and heavy distillates, and 4% asphalt[1]. Asphalt consists of four fractions (saturates, aromatic, resins, and asphaltenes) and it can be separated by column chromatography (including the ASTI D4124[2], solvent fractionation[3], and SARA[4] methods) or planar chromatography (such as thin-layer chromatotron[5], and thin-layer chromatography interfacial with flame ionization detector (TLC-FID)[6]).

Asphaltene Particles in Fossil Fuel Exploration, Recovery, Refining, and Production Processes, Edited by M.K. Sharma and T.F. Yen, Plenum Press, New York, 1994

193

Asphalt is referred as a colloidal, homogeneous system because resins act as peptizing agents which can be adsorbed on the particles of asphaltenes and enable them to dissolve well in gas oil (saturates and aromatic) (Figure 1)[7]. Due to these characteristics, asphalt possesses some unique and excellent properties. It is thermoplastic, adhesive toward most other materials, and resistant to water, acids, and alkalis. Thus, asphalt can be applied in the diverse fields of hydraulics (dam facing, canal and pond sealers, etc.), recreation (substratum for artificial surfaces, tennis courts, running tracks, etc.), agriculture (mulches, underground barriers, stockyard paving, etc.), transportation (railroad ballast treatment and roadbeds), and metals (ore leaching, ore and coal briquette, etc.)[8]. Besides that, asphalt has also been utilized as fixation agent to remediate underground soil contamination by hydrocarbons or hazardous materials[9].

Asphalt can be classified into so, so-gel, and gel types by traditional rheologic indexes such as complex flow, asphalt aging index, plastic flow, penetration index, ductility at 25 °C, etc., and the correlation of asphaltene aromaticity with H/C ratio[10]. In general, so and so-gel type asphalt can be used on paving, while gel type asphalt can be used on roofing, and metal coating. In industry, the majority of gel type of asphalt were produced by air-blowing so, or so-gel type of asphalt under high temperature and high pressure[11]. Recently, the public has been seriously concerned by environmental impacts on human health, so that many regulations relating to air pollution, water pollution, and subsurface soil-water contamination are being established by EPA and other agencies. Asphalt emulsion is one of the alternatives to avoid volatile toxic compounds emitted in to the air during paving, curing, roofing, and metal coating processes that are normally operated at high temperatures. In this research, we are attempting to invent a new method to obtain a sol-gel or gel type asphalt emulsion by dissolving so or sol-gel type of asphalt in water with a surfactant under ultrasound in order to avoid emission problems and save energy. The detailed procedures will be discussed in the following sections.

EXPERIMENTAL

Figure 2 shows that sol-gel, or gel type asphalt can dissolve in water when an emulsifier (surfactant) is added, ferrous sulfate is added as a catalyst, and finally hydrogen peroxide is added as an oxidant under sonication to obtain either an encapsulating asphalt emulsion or a coating asphalt emulsion (depending on the reaction time, oxidant concentration, sample solution concentration, energy input, etc.). Here we selected AAD-1 asphalt (from California coast) as the sample because this asphalt had a higher asphaltene content, and could be on the margin of so-gel and gel type asphalt[12]. Also, anionic surfactant, sodium di(2-

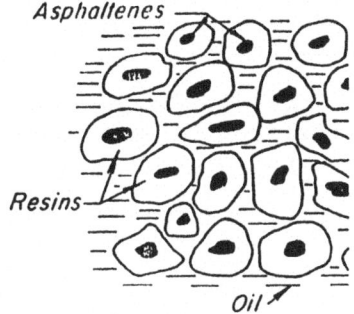

Figure 1. The compositions of asphalt

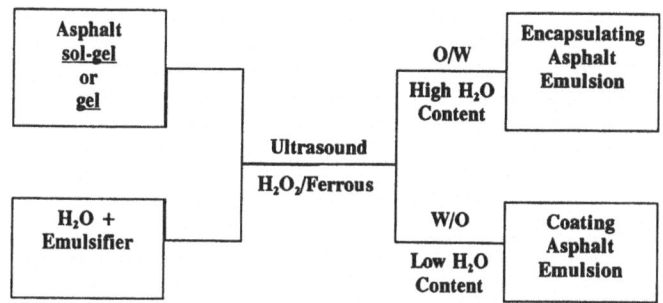

Figure 2. The scheme of emulsion polymerization

Aerosol OT: sodium di(2-ethylhexyl) sulphosuccinate

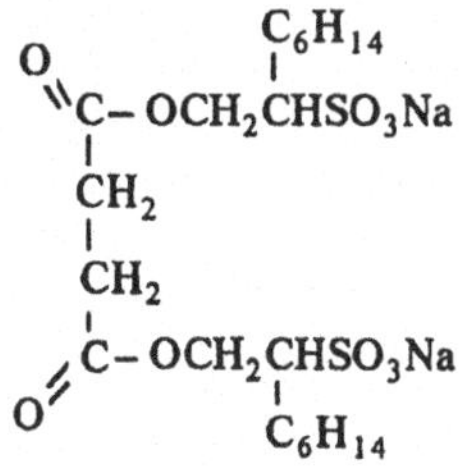

$$
\begin{array}{l}
\quad\quad\quad\quad\quad C_6H_{14} \\
O \quad\quad\quad\quad | \\
\;\diagdown C- OCH_2CHSO_3Na \\
\quad | \\
\quad CH_2 \\
\quad | \\
\quad CH_2 \\
\quad | \\
\;\diagup C- OCH_2CHSO_3Na \\
O \quad\quad\quad\quad | \\
\quad\quad\quad\quad\quad C_6H_{14}
\end{array}
$$

Figure 3. The structure of anionic surfactant - Aerosol OT

ethylhexyl) sulphosuccinate (Aerosol OT) was selected as an emulsifier, and Figure 3 illustrates its structure. Hydrogen peroxide and potassium superoxide were chosen as oxidants in these experiments.

First, 5 grams of AAD-1 were dissolved in 5 grams of toluene with a 150 milliliter beaker; then 100 grams of water was introduced to the solution. After that, 0.0275 grams of potassium superoxide was added in four portions within 2 hours during sonication where the mixture (water plus the sample solution) was agitated by a magnetic bar at 1000 rpm. A VC-50 Ultrasonic Processor was used in the experiment and the frequency and energy output were 20 KHz and 50 Watts. The total amount of potassium superoxide used in this experiment was 0.11 grams (0.1 weight percentage of total mixture solution). In order to prove that oxidant concentration and surfactant can affect polymerization processes of asphalt emulsion during sonication, the following two experiments were repeated with the same materials and systems as the above experiment. One used more emulsifier-Aerosol OT (0.22 grams (0.2% wt.)); however, the other used more oxidant, hydrogen peroxide (0.44 grams (0.4% wt.)).

Second, with the same ultrasound condition and mixture, 0.22 grams of Aerosol OT (0.2% wt.), 0.22 grams of hydrogen peroxide (0.2% wt.), and 0.55 grams of ferrous sulfate (0.5% wt.) were added to repeat the above experiment. Here the reaction time was reduced from 2 hours to one hour, and hydrogen peroxide was divided into four portions to be added every 15 minutes during the sonication processes.

Finally, another experiment was done by using the same conditions and materials, except that 0.22 grams of the emulsifier-Aerosol OT (0.2% wt.) were used only when no oxidants were added. The reaction time was 1 hour.

RESULTS AND DISCUSSIONS

In order to overcome the high temperature susceptibility for special usages such as roofing and coating, a mixture of gas oil and resin can be converted into resin and asphaltene, which shall change asphalt colloidal properties from so-gel type to gel type through emulsion polymerization under ultrasound. The traditional blown asphalt were produced from steam or vacuum refinery systems in which asphalt residues were blown by air at 450 to 550 °C[13-14]. Fortunately, ultrasound can create cavitation conditions for the formation of radicals at the localized high temperature (5075 K) and high pressure (75000 psi) microenvironments[15-16]. Another concern is that the emission of toxic organic compounds such as benzene, toluene, etc., during the operating precess will be eliminated by using asphalt emulsion. Figure 4a illustrates the detailed procedures of manufacturing an asphalt emulsion when asphalt and water with emulsifier are pumped together through a

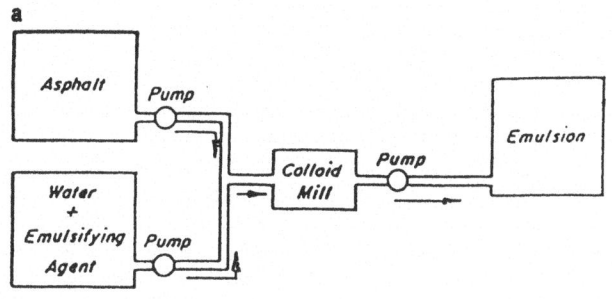

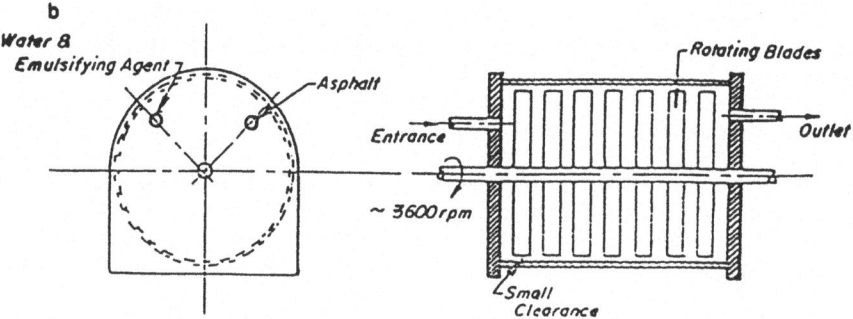

Figure 4. Simplified flow diagram (TOP) and schematic representation of colloidal mill

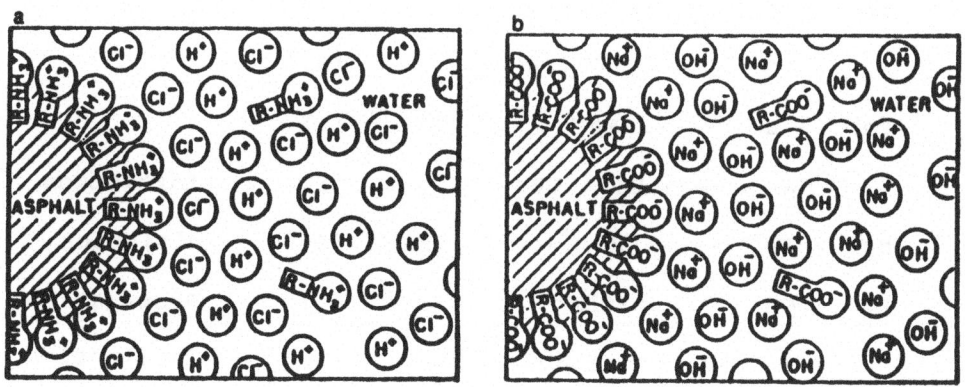

Figure 5. Emulsified asphalt droplet (a) cationic emulsion (b) anionic emulsion. (After Mertens & Wright)

colloidal mill. Here both the asphalt and the water
emulsifier mixed solution are heated up to 180 °F before
being pumped to the colloidal mill, and the plan of
colloidal mill is presented on Figure 4b. The mechanisms of
emulsion formed are that under high revolutions (3600 rpm)
emulsifiers can be easily adsorbed on asphalt particles,
then enhanced asphalt particles can be dissolved in the
water by shearing stress. In ultrasonic systems it is easy
to form an emulsion by acoustic streaming which can avoid
asphalt aggregation, and thus yield sufficient energy for
forming asphalt emulsion. In this research we select
Aerosol OT as an anionic emulsifier, and after sonication an
anionic asphalt emulsion will be formed. Figure 5 indicates
structures of an anionic asphalt emulsion by Mertens and
Wright.[17]

The three steps for emulsion polymerization are
initiation, propagation, and termination[17]. Hydrogen
peroxide with a ferrous sulfate system are selected because
ferrous ions can act as catalysts as Equations 1-4 if the
hydrogen peroxide concentration can be controlled. Those
reactions are already proven by Haber and Weiss and other
researchers[18-21]. Under ultrasound, hydroxyl free radicals
are formed by initiation; then through propagation and
termination a small molecular weight of gas oil or resin
will be converted into resin or asphaltene see Equations (5)
to (9).

$$H_2O_2 + Fe^{++} \longrightarrow HO^{\cdot} + OH^- + Fe^{+++} \quad (1)$$

If $[Fe^{++}] > [H_2O_2]$, then

$$Fe^{++} + HO^{\cdot} \longrightarrow Fe^{+++} + OH^- \quad\quad (2)$$

If $[H_2O_2] > [Fe^{++}]$, then

$$H_2O_2 + HO^{\cdot} \longrightarrow H_2O + HO_2^{\cdot} \quad\quad (3)$$

$$Fe^{+++} + HO_2^{\cdot} \longrightarrow Fe^{++} + O_2 + H^+ \quad\quad (4)$$

$$H_2O \longrightarrow H^{\cdot} + OH^{\cdot} \quad\quad (5) \quad\quad \textbf{Initiation}$$

$$H_2O_2 + Fe^{++} \longrightarrow HO^{\cdot} + OH^- + Fe^{+++} \quad (6) \quad\quad \textbf{Initiation}$$

$$RR' + OH^{\cdot} \longrightarrow HO\overset{*}{R}R' \quad\quad (7) \quad\quad \textbf{Propagation}$$

$$HO\overset{*}{R}R' + HO\overset{*}{R}R' \longrightarrow HORR'R'ROH \quad\quad (8) \quad\quad \textbf{Termination}$$

or

$$HO\overset{*}{R}R' + OH^{\cdot} \longrightarrow ORR' + H_2O \quad\quad (9) \quad\quad \textbf{Termination}$$

where R is the methyl group, R' is the aromatic group.

Three different systems are compared to each other and
with the original AAD-1 asphalt shown in Figure 6. These
results indicate that asphaltene contents are increased up

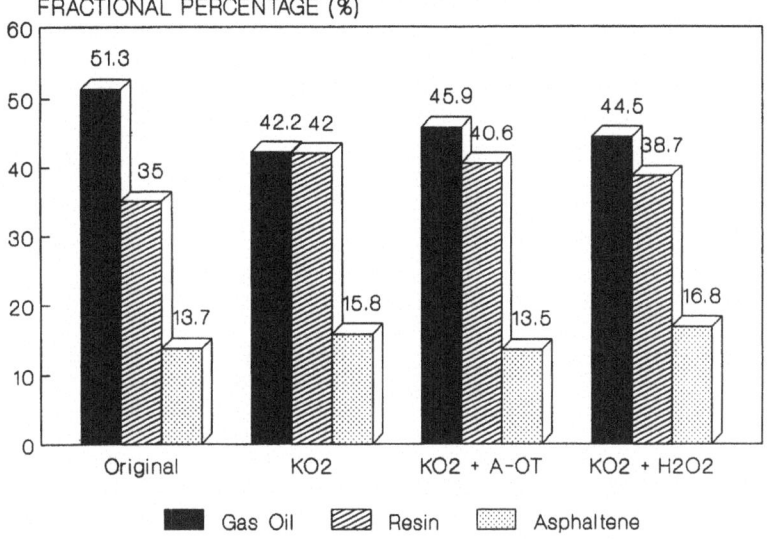

FRACTIONAL PERCENTAGE (%)

KO2 - 0.1%,   H2O2 - 0.4%
A-OT - 0.2%,   Time - 2 hours

Figure 6. The comparison of AAD-1 asphalt contents by adding
surfactant and oxidant under ultrasound

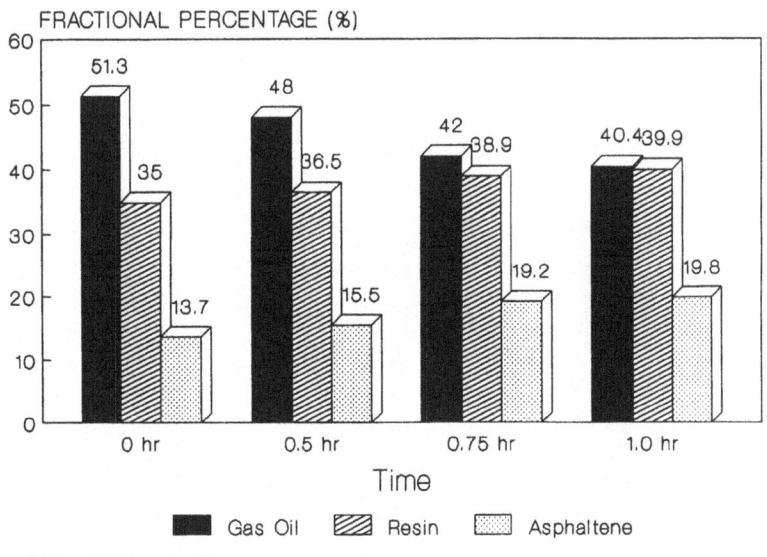

FRACTIONAL PERCENTAGE (%)

A-OT - 0.2%,   H2O2 - 0.2%
FeSO4 - 0.5%

Figure 7. The comparison of AAD-1 asphalt contents at
different time by adding surfactant, oxidant,
and catalyst under ultrasound

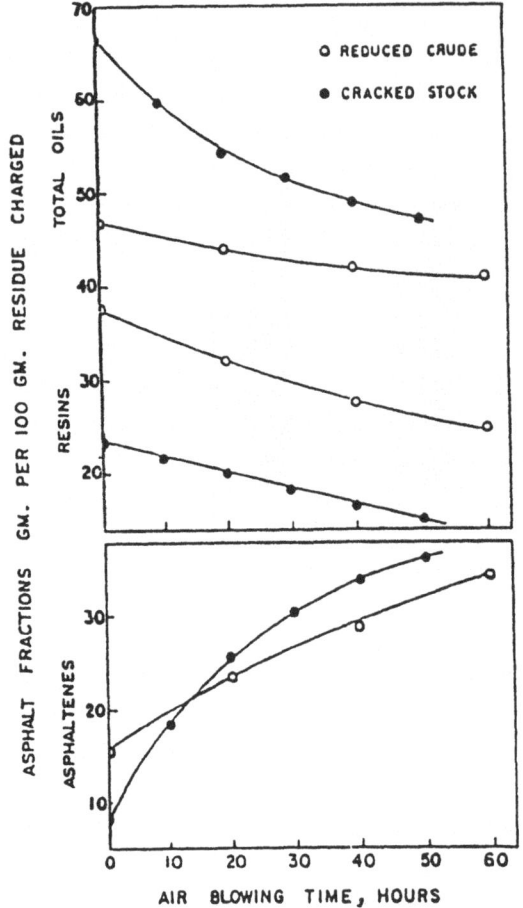

Figure 8. The change in gas oil, resin, asphaltene content at different time through air blowing process

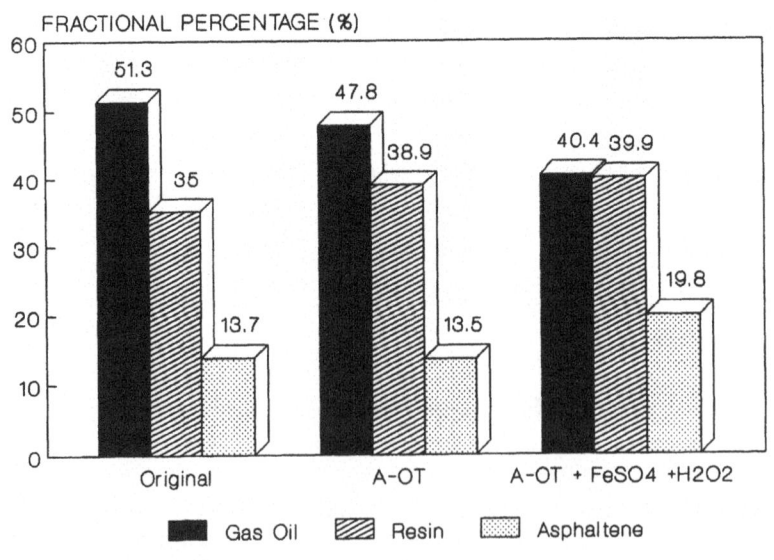

A-OT - 0.2%, H2O2 - 0.2%
FeSO4 - 0.5%, Time - 1 hour

Figure 9. The comparison of AAD-1 asphalt contents for different systems under ultrasound

to 2% or 3% when 1% by weight potassium superoxide is added; or potassium superoxide 0.1% and hydrogen peroxide 0.4% by weight. But if we replace hydrogen peroxide 0.4% by weight with Aerosol OT 0.2% by weight, then asphaltene contents remain the same. There exists a possibility that asphaltene content is without any noticeable change under ultrasound. The explanation is that when the Aerosol OT is dissolved in the water it will form sodium ions to neutralize the hydroxyl ions generated by dissolving potassium superoxide into water. In this manner, neutralization will reduce the formation of hydroxyl free radicals. Figure 7 indicates the success of the emulsion polymerization under ultrasound, that the asphaltene contents in AAD-1 were increased up to 6% by weight within 1 hour by adding Aerosol OT 0.2%, hydrogen peroxide 0.2%, and ferrous sulfate 0.5% by weight. By comparing with the traditional air-blown method shown on Figure 8[22], the efficiency of emulsion polymerization under ultrasound is about six times higher than the traditional blown method (less than 1% asphaltene increased within 1 hour). Finally, two systems, one adding emulsifier Aerosol OT 0.2% by weight only, the other adding emulsifier Aerosol OT 0.2% with ferrous salt 0.5% and hydrogen peroxide 0.2% by weight, are compared with original AAD-1 asphalt. The result shown in Figure 9 indicates the function of the emulsifier is to enhance the dissolving of the asphalt in the water only. The asphalt content can be increased during short period if oxidant and catalyst are added in the system.

CONCLUSIONS

Blown asphalt can be made by emulsion polymerization under ultrasound. The function of an emulsifier is to stabilize the emulsion system only. Hydrogen peroxide and potassium superoxide are good oxidants which can generate hydroxyl free radicals by sonication, propagation, and termination to increase asphaltene contents. Ferrous salt can be used as a catalyst to enhance the asphaltene conversion process. This emulsion polymerization under ultrasound method combines two processes into one single process that forms blown asphalt, and thus to an asphalt emulsion. Through this method different types of asphalt emulsions can be obtained for various usages.

REFERENCES

1. Izatt, J., "Asphalt", Encyclopedia of Chemical Processing and Design (edited by Mckett, J.J.)..3:421, Marcel Dekker, New York (1986).

2. ASTM D4124, "Standard Test Method for Separation of Asphalt into Four Fractions", American Society for Testing and Materials, Philadelphia, Pa., (1988).

3. Schwager, J. and Yen, T. F. "Coal Liquefaction Products from Major Demonstration Processes I. Separation and Analysis", Fuel, 57(2), 100-104 (1978).

4. Altgelt, K. H., Jewell, D. M., Latham, D. R. and Selucky, M. L.; "Chromatography in petroleum analysis", New York and Basel:Marcel Dekker Inc. In Chromatographic Science Series. ed. K. H. Altgelt and T., p.194-196 (1979).

5. Lian, H. J., Lee, C., Wang, Y. Y. and Yen, T. F.; "Characterization of Asphalt with the Preparative Chromatotron", Journal of Planar Chromatography, 5, 263-266 (1992).

6. Lian, H. J., Lee, C. and Yen, T. F.; "Fractionation of Asphalt by Thin-Layer Chromatography interfacial with Flame Ionization Detector (TLC-FID) and Characterization by FTIR", Fine Particle Society Conference, Las Vegas, July, (1992).

7. Monismith, C. L.; "Asphalt Paving Mixtures:Design, Construction, and Performance", p.121, (1984).

8. The Asphalt Institute, "Introduction to Asphalt:Manual Series No. 5. English Edition", College Park, MD, (1986).

9. Preston, R. L., and Testa, S. M. "Permanent Fixation of Petroleum-Contaminated Soils", Contaminated Soils, p.129-132 (1990).

10. Barth E. J.; "Asphalt:Science and Technology", chap 4., p.252, 1962.

11. Barth, E. J.; "Asphalt:Science and Technology", chap 6., p.386 (1962).

12. Strategic Highway Research Program:Asphalt Contractors Meeting Notes, attachment 4, Denver, Colorado, February 2-3 (1989).

13. Fauber, E. M., "Air Brown Asphalt Pitch Composition", U.S. Patent 3462359 (1969).

14. Goodrich, J. E.; "Asphalt Composition for Air-Blowing", U.S. Patent 4338137 (1982).

15. Frederick, J. R.; "Ultrasonic Engineering", John Wiley and Sons, New York (1965).

16. Flient, E. B., and Suslick, K. S.; "The Temperature of Cavitation", Science 253:1397-1399, 1991.

17. Wright, J. R. and Mertens, E. W.; Theoretical and Practical Aspects of Cationic Asphalt Emulsions", Proceedings, Highway Research Board, vol. 28 (1959).

18. Bovey, F. A., Kolthoff, I. M., Medalia, A. I. and Meehan, E. J.; "Emulsion Polymerization", Interscience Publishers, Inc., New York, chapter III (1955).

19. Haber, F. and Weiss, J.; Naturwissenschaften, 20, 948 (1932).

20. Medalia, A. I., and Kolthoff, I. M.; "The reaction between hydrogen peroxide and ferrous iron", J. Polymer Sci., 4, 377 (1949).

21. Weiss J.; "The reaction between hydrogen peroxide and ferrous iron, in Advances in Catalysis", Academic Press, New York, Vol. IV, p. 343, 1952.

22. Baxendale, J. H.; "The reaction between hydrogen peroxide and ferrous iron", ibid., p. 31.

23. Earth, E. J.; "Asphalt:Science and Technology", chap 6.,p.394, (1962).

AN IMPROVED MOLECULAR THERMODYNAMIC MODEL

OF ASPHALTENE EQUILIBRIA

V. A. Kamath, M. G. Kakade and G. D. Sharma

Petroleum Development Laboratory
University of Alaska Fairbanks
Fairbanks, AK 99775-1260

ABSTRACT

In this study, a molecular thermodynamic model was developed to represent asphaltene equilibria and to predict the amount of asphaltene precipitation that would occur from a reservoir oil under influence of a miscible solvent or immiscible gas. The model treats the asphaltenes to exist in the crude in a large range of molecular weights represented by a normal distribution function. The properties of each asphaltene pseudo-component such as solubility parameter and molar volume are obtained based on their molecular weights at given pressure and temperature. Scott-Magat theory along with binary interaction parameter between asphaltene free liquid phase (solvent) and asphaltenes are used to represent asphaltene solid-liquid equilibria and to predict the degree of asphaltene precipitation, the molar distribution of asphaltene pseudo-components in equilibrium solid and liquid phases at various pressures, temperatures and solvent-oil compositions. The model is coupled with Peng-Robinson equation of state for vapor-liquid equilibria calculations. The model uses an iterative Newton-Raphson scheme to obtain solution. The model has four parameters which are determined by fitting the model to experimental asphaltene precipitation data for tank West Sak oil-solvent mixtures. The model is used to predict asphaltene precipitation for CO_2-West Sak oil mixtures at various pressures.

Asphaltene Particles in Fossil Fuel Exploration, Recovery, Refining, and Production
Processes, Edited by M.K. Sharma and T.F. Yen, Plenum Press, New York, 1994

205

INTRODUCTION

Many field cases and laboratory studies have reported precipitation and deposition of asphaltenes during recovery of asphaltic crudes by immiscible or miscible gas flooding operations[1-4]. In many field cases, asphaltene deposition in reservoirs have severely reduced well injectivities and productivities[5,6]. Asphaltene precipitation have also caused technical problems in surface separation and oil upgrading[7]. Other field conditions conducive to asphaltene precipitation include: natural depletion, gas-lift operations, caustic flooding and well treatments by acidization[8,9]. Industry is looking for ways to control asphaltene deposition and economic ways to remedy the problem. The development of a predictive technique is crucial to the solution of the problem.

In miscible flooding, enriched hydrocarbon gases, LPG's or CO_2 (termed as solvents) are injected into the reservoir to reduce oil viscosity and interfacial tension, to provide drive mechanism and to achieve (dynamic or direct contact) miscibility with reservoir oil thus resulting into very high displacement efficiency and ultimate oil recovery. In such processes, however, the miscible solvents can also cause precipitation of asphaltenes by altering the resin to asphaltene ratio of the crude. The amount of asphaltene precipitation depends upon the reservoir oil composition (the original asphaltene and resin content of the crude), the solvent composition, the pressure and temperature conditions and the phase behavior of solvent-oil mixtures. Many of the previous studies have provided fundamental understanding of the cause and mechanism of asphaltene flocculation[7,10]; chemical and physical properties, and adsorption characteristics of asphaltenes[11,12].

Asphaltene flocculation in porous media and it's interactions with rock and fluids are complex dynamic phenomena. The in-situ precipitated asphaltenes may aggregate and adsorb on the rock surfaces altering the wettability of rock. Clay minerals in reservoir formations can strongly adsorb asphaltic material and such adsorption has been shown to alter the wettability of the rock to a more oil-wet condition[11-13]. Also, deposition of asphaltic materials near the wellbore may cause pore plugging, and permeability reduction and thus block the flow of oil. Recently, Yang[14] conducted experimental work on the effect of asphaltene deposition on permeability reduction, oil-water relative permeability curves and on the overall oil recovery during dynamic displacement of oil by water.

REVIEW OF ASPHALTENE EQUILIBRIA MODELING

Two distinct approaches have emerged in formulating asphaltene precipitation thermodynamic model. These are: (1) thermodynamic molecular approach, and (2) thermodynamic colloidal approach. In the first approach (molecular), asphaltenes molecules are considered to be dissolved in oil in a true liquid state and may precipitate as a result of

changing thermodynamic conditions such as temperature,
pressure and composition. Thus, asphaltene precipitation is
considered to be thermodynamically reversible. The
dissolution of asphaltenes in oil takes longer time than
precipitation. The complete dissolution of some asphaltenes
in organic solvents like toluene and benzene supports this
assumption. In the second approach (colloidal), asphaltenes
are considered to colloidally suspended in crude oil which
are stabilized by the adsorbed resin molecules. When the
adsorbed resins are dissolved into solution, the asphaltene
particles may aggregate mechanically or by electrostatic
attraction.

Thermodynamic Molecular (Solubility) Approach

The different asphaltene equilibria models emerged
within the thermodynamic molecular (solubility) approach
type, can be classified under two categories based on the
treatment of asphaltenes as: (i) monodispersed (homogeneous
or single component) polymeric molecules and, (ii)
polydispersed (or heterogeneous or multi-component)
polymeric molecules.

Models presented by Fussel[15], Hirschberg et al.[16],
Burke et al.[17] and Kamath et. al.[18] fall under the
monodispersed type. These models are based on the treatment
of asphaltenes as a single pseudo component having an
average molecular weight, molar volume and solubility
parameter. With some modifications these models combine the
Flory-Huggins polymer solution theory with the equation of
state vapor/liquid equilibria calculations to predict the
solubility parameter and molar volume of the asphaltene
free-liquid phase and the amount of asphaltene
precipitation. Kamath et al.[18] applied the homogeneous model
to predict the degree of asphaltene precipitation for
various solvent-West Sak oil mixtures and found that
homogeneous models have several limitations. These include:
(1) insufficient representation of asphaltenes within oil by
a single component and (2) limited range of application of
the model since molar volume and solubility parameter of
asphaltenes treated as independent of pressure and
temperature. Recently, Novasad and Costain[19] adopted
Hirschberg's molecular solubility model, but introduced
association parameter to represent resin-asphaltene
interaction. They were able to successfully correlate the
model with the experimental asphaltene precipitation data,
but recommended that asphaltenes be described as continuous
(polydispersed) phase to improve the accuracy of
predictions.

Kawanaka et. al.[20] were the first to represent
asphaltenes as continuous polydispersed molecules having
molecular weight distribution. Their model uses Scott-Mogat
theory to represent asphaltene equilibria and represents an
improvement over the homogeneous type models. The model
developed in this paper also treats asphaltenes as
multicomponent system, but differs from Kawanaka et al.'s
model in many other ways.

Thermodynamic Colloidal Approach

This approach assumes that asphaltenes exist in the oil as solid particles in suspension, stabilized by resins (peptizing agents) adsorbed on their surface. The intermolecular repulsive forces between neutral aromatic resin molecules adsorbed on the polar asphaltene particles prevent them from flocculating. When the concentration of resins falls below the critical resin concentration (due to addition of paraffinic solvents), asphaltene particles may aggregate due to neutralization of their weak repulsive forces by mechanical (agitation) or electrical (opposing stream potential) means. Leontaritis and Mansoori[21] considering the above factors developed a thermodynamic-colloidal model capable of predicting the onset of asphaltene flocculation. Their model contains two parts: a static model to determine the chemical potential of resin based on the Flory-Huggins theory and a dynamic model to determine the stream potential for asphaltene to precipitate.

DESCRIPTION OF MULTICOMPONENT ASPHALTNENE MODEL

Since asphaltenes exists in crude oil in large range of molecular weights, for proper representation of asphaltene equilibria we considered multi-component asphaltene modeling approach. Figure 1 shows a schematic representation of various phases and fractions of oil and oil-solvent mixtures for three situations: (a) asphaltenes in reservoir oil, (b) asphaltene precipitated by addition of solvents to tank oil, (c) asphaltenes precipitated by addition of solvents to live oil. In case (a), the live oil (LO) is considered to be a mixture of N_T components of which, N_F are asphaltene free components and N_A are asphaltene pseudocomponents. The live oil is made up of solution gas (SG) which contains asphaltene free lighter hydrocarbons which are dissolved and tank oil (TO) which is further divided into asphaltene free oil (B) and dissolved asphaltene components (A_i, i=1, N_A). In case (b), upon addition of certain proportion of solvent (SOL) to the tank oil (TO), an overall oil-solvent mixture (TO_{mix}) is formed which splits into vapor (G) and liquid (L_{mix}) phases according to VLE calculations. The L_{mix} phase instantaneoulsy splits into an equilibrium solid (S) phase which contains only precipitated asphaltene components and an equilibrium liquid (L) phase according to solid-liquid equilibrium calculations. Finally, the L phase consists of asphaltene free liquid (B) phase and dissolved asphaltene fractions (A_i). The case (c) is similar to case (b) except solvents are added to live oil (LO) rather than tank oil (TO).

Discretization of Continuous Asphaltene Molecular Weight Distribution Function

If we consider the asphaltenes in crude oil in a large range of molecular weights described a continuous normal distribution function, $F(M_A)$ given by:

A. **Asphaltenes in Live Reservoir Oil**

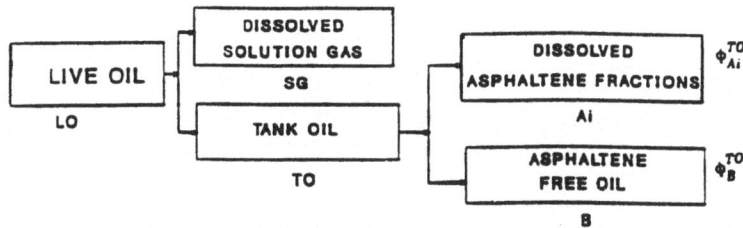

B. **Asphaltene Precipitation from Addition of Solvents to Tank Oil**

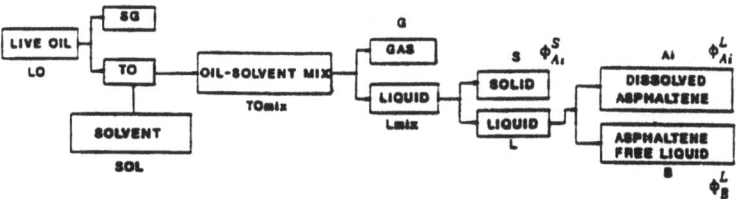

C. **Asphaltene Precipitation from Addition of Solvents to Live Oil**

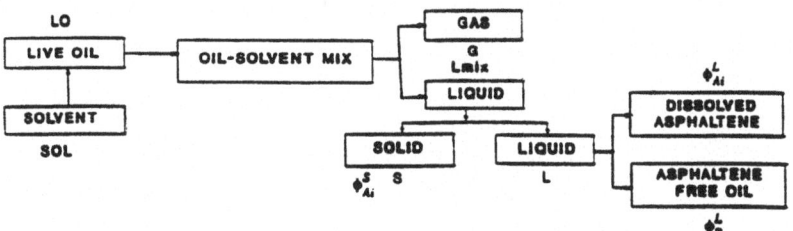

Figure 1. Schematic Representation of Various Fractions and Phases of Oil and Oil-Solvent Mixtures

$$F(M_A) = \frac{1}{\sigma\sqrt{2\pi}} \exp\left(-\frac{\xi^2}{2}\right) \tag{1}$$

where,

$$\xi = \frac{M_A - \overline{M_A}}{\sigma} \tag{2}$$

M_A is the molecular weight of asphaltenes treated as a continous variable, $\overline{M_A}$ is the average molecular weight of asphaltenes and σ is standard deviation of the normal distribution function.

If we discretize the continuous normal distribution function as shown in Figure 2 such that the area under the two curves are same, then the discretized distribution is characterized by N_A equal intervals of ΔM_A and the initial value MA_o and the final value M_{AF}. Theoretically, M_{Ao} tends to $-\infty$ and M_{AF} tends to $+\infty$. But for practical purposes, it is sufficient to approximate the span of molecular weight to 6σ since it causes only 0.26% error.

Therefore,

$$M_{AO} = \overline{M_A} - 3\sigma \tag{3}$$

$$M_{AF} = \overline{M_A} + 3\sigma \tag{4}$$

$$\int_{-\infty}^{+\infty} F(M_A)\, dM_A \approx \sum_{i=1}^{N_A} F(M_{Ai})\, \Delta M_A = 0.9974 \approx 1 \tag{5}$$

Thus, the molecular weight of ith component of asphaltenes is given by:

$$M_{Ai} = M_{Ao} + \frac{3\sigma}{N_A}[2i-1], \quad i = 1, N_A \tag{6}$$

CONTINUOUS

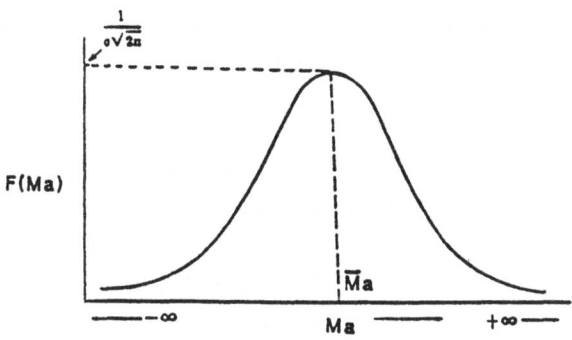

DISCRETIZED

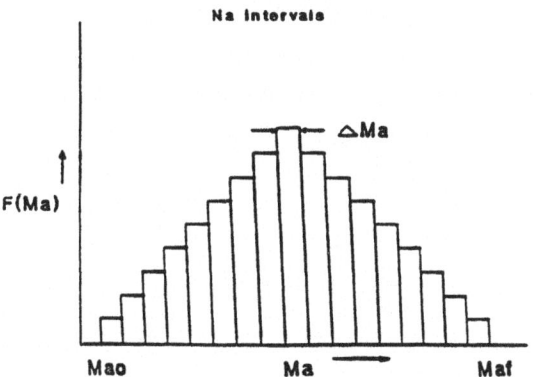

Figure 2. Discretization of a Continuous Normal Distribution Function of Asphaltenes Fractions

Also,

$$\xi_i = \frac{3}{N_A}[2i-1-N_A], \quad i = 1, N_A \tag{7}$$

If X_{AT}^{TO} is the total mole fractions of asphaltenes in the tank oil, then

$$X_{AT}^{TO} = \sum_{i=1}^{N_A} X_{Ai}^{TO} \tag{8}$$

and

$$\overline{M_A^{TO}} = \frac{1}{X_{AT}^{TO}}\sum_{i=1}^{N_A} M_{Ai}X_{Ai}^{TO} \tag{9}$$

The mole fraction of ith asphaltene component is then obtained by

$$X_{Ai}^{TO} = \frac{3\sqrt{2}\,X_{AT}^{TO}}{\sqrt{\pi}\,N_A}\exp(-0.5\xi_i^2), \quad i=1, N_A \tag{10}$$

Asphaltene Pseudocomponent Property Estimation

The molar volume of each asphaltene pseudocomponent is given by:

$$V_{Ai} = \frac{M_{Ai}}{\rho_A} \tag{11}$$

where ρ_A is the average density of the asphaltenes.

Twu's[22] critical property correlation is used to compute critical temperature (T_{ci}) and the normal boiling point (T_{bi}) for ith asphaltene component based on the molecular weight, (M_{Ai}) Kistyakowsky's equation[23] is then used to compute the heat of vaporization given by:

$$(\Delta H_i)_T = (\Delta H_i)_{TB}\left(\frac{T_{ci}-T}{T_{ci}-T_{bi}}\right) \tag{12}$$

where

$$(\Delta H_i)_{TB} = 1.014 \left[T_{bi}(8.75 + 4.571\log(T_{bi})) \right] \tag{13}$$

where T_{ci}, T_{bi} and T are in (°K) and $(\Delta H_i)_T$ is in (cal/gmole). The solubility parameter for each asphaltene component is then given by:

$$\delta_{Ai} = \left[\frac{(\Delta H_i)_T - RT}{V_{Ai}} \right]^{0.5} \tag{14}$$

Asphaltene Solid-Liquid Equilibria

Using the Scott-Magat theory[24] to represent solid-liquid equilibria for asphaltenes as proposed by Kawanaka et al., the solid-liquid equilibrium constant for the ith asphaltene component, (K_{Ai}) is given by:

$$K_{Ai} = \frac{X_{Ai}^S}{X_{Ai}^L} = \frac{\phi_{Ai}^S}{\phi_{Ai}^L} \cdot \frac{V_A}{V_L} = \frac{V_A}{V_L} \exp\left[(\frac{V_{Ai}}{V_A} - \frac{V_{Ai}}{V_B}) \phi_B + f \frac{V_{Ai}}{V_B} \phi_B^2 \right] \tag{15}$$

where superscripts S and L refer to equilibrium solid and liquid phases respectively, X_{Ai} is the mole fraction and Φ_{Ai} is the volume fraction of ith component asphaltenes, V_A and V_B are the molar volumes of asphaltenes and asphaltene-free liquid phase respectively given by:

$$V_A = \sum_{i=1}^{N_A} (V_{Ai} X_{Ai}^{Lmix}) / X_{AT}^{Lmix} \tag{16}$$

$$V_B = (V_{Lmix} - X_{AT}^{Lmix} \cdot V_A) / (1 - X_{AT}^{Lmix}) \tag{17}$$

where V_{Lmix}, X_{Ai}^{Lmix} and X_{AT}^{Lmix} are molar volume, mole fraction of ith asphaltene component and total asphaltene mole fraction in Lmix phase respectively. V_L is the molar volume of equilibrium liquid phase. ϕ_B is the volume fraction of asphaltene free components in Lmix phase which is obtained as:

$$\phi_B = 1 - \phi_A = 1 - \sum_{i=1}^{N_A} \phi_{Ai}^{Lmix} \qquad (18)$$

and

$$\phi_{Ai}^{Lmix} = \frac{X_{Ai}^{Lmix} V_{Ai}}{V_{Lmix}} \qquad (19)$$

The parameter f in equation (15) is defined as:

$$f = \frac{1}{r} + \frac{V_B}{R_T} [(\delta_A - \delta_B)^2 + 2K_{AB}\delta_A\delta_B] \qquad (20)$$

where r is coordination number (between 3 and 4), K_{AB} is the binary interaction parameter between asphaltenes and asphaltene free components, δ_A and δ_B are the average solubility parameters for asphaltenes and asphaltene free components given by:

$$\delta_A = \frac{1}{\phi_A} [\sum \phi_{AI}^{Lmix} \delta_{Ai}] \qquad (21)$$

$$\delta_B = (\delta_{Lmix} - \delta_A\phi_A)/\phi_B \qquad (22)$$

In equation (20), the first term comes from entropy of mixing whereas the second term comes from the enthalpy of mixing.

The interaction parameter K_{AB} is treated as the linear function of the average molecular weight of asphaltene free components in the L_{mix} phase.

$$K_{AB} = a' + b' \overline{M}_B^{Lmix} \qquad (23)$$

where a' and b' are parameters obtained by fitting experimental asphaltene precipitation data for tank oil-solvent mixtures. Thus, this interaction parameter incorporates effect of solvent/oil ratio and solvent molecular weight on the degree of asphaltene precipitation.

Vapor-Liquid Equilibria

In order to determine the equilibrium L_{mix} phase composition (X_i^{Lmix}), molar volume (V_{Lmix}), and solubility parameter (δ_{Lmix}), vapor-liquid equilibria calculations for a

214

given oil-solvent mixture (Z_i^{Lomix} or Z_i^{Tomix}) at given pressure and temperature are performed using the Peng-Robinson equation of state as described in Kamath et al.[18].

Stepwise Model Calculations and Regression of Experimental Data

The model is capable of predicting the amount of asphaltene precipitation, the molar distribution of various asphaltene components in the equilibrium solid (S) and liquid (L) phases and the equilibrium constant for each asphaltene component (K_{Ai}) for a given oil-solvent mixture at given pressure and temperatures. The input data to the model includes: the overall composition of the oil, composition of solvent, solvent to oil ratio, pressure, temperature and values for four model parameters namely: a', b', σ and β (the ratio of total mole fraction of asphaltenes to mole fraction of plus fraction in the tank oil). The model can be used in predictive mode or regression mode. In the regression mode, the experimental asphaltene precipitation data for tank oil/solvent mixtures for various solvent/oil ratios are fitted to obtain the four model parameters. The stepwise procedure in predictive mode is given below:

Step 1)   The PVT data for the oil is initially used to tune the Equation of State parameters.

Step 2)   The VLE calculations are then performed at given P, T and overall oil-solvent mixture composition to determine flashed equilibrium liquid phase properties (X_i^{Lmix}, V_{Lmix} and δ_{Lmix}).

Step 3)   The mole fractions (X_{Ai}^{TO} or X_{Ai}^{LO}, X_{Ai}^{Lmix}) and molecular weights (M_{Ai}) of each asphaltene components in the tank oil (or live oil) and the oil-solvent mixture (L_{mix}) are obtained using equations (1) through (10). Note that superscript TO can be replaced by LO or L_{mix} in equations (8-10). Also note that:

$$X_{AT}^{j} = \beta X_{+}^{j} \quad j = TO, \ LO \ or \ Lmix \qquad (24)$$

Step 4)   The molar volume (V_{Ai}) and solubility parameter (δ_{Ai}) are calculated using equations (11) through (14). Also, V_A, V_B, Φ_A, δ_A, δ_B are calculated using equations 16, 17, 18, 19, 21, and 22 respectively.

Step 5)   The ratio $\phi_{Ai}^{s}/\phi_{Ai}^{L}$; and K_{Ai} are obtained using equations (23), (20), and (15), in that order.

Step 6)   Newton Raphson iterative technique is used to
          calculate molar volume of equilibrium liquid phase
          (V_L) using $V_L = 0.95\ V_{Lmix}$ as initial guess. The
          function $f(V_L)$ is defined and calculated as:

$$f(V_L) = V_A - \sum_{i=1}^{N_A} \frac{V_{Ai} X_{Ai}^{Lmix}}{\left[\left(\dfrac{V_{Lmix}-V_A}{V_L-V_A}\right) + \left(\dfrac{V_L-V_{Lmix}}{V_L-V_A}\right)K_{Ai}\right]} \qquad (25)$$

          The new value of V_L is then obtained as:

$$V_L(new) = V_L(old) - \frac{f(V_L)}{f'(V_L)} \qquad (26)$$

          where $f'(V_L)$ is the derivative of $f(V_L)$ with
          respect to V_L. These calculations are repeated
          till the following convergence criteria is
          satisfied:

$$\left(\frac{f(V_L)}{F'(V_L)\ V_L}\right) \le 0.001 \qquad (27)$$

Step 7)   By considering the material balance for each
          asphaltene component between L_{mix}, L and S phases
          the following expressions are derived and used to
          calculation mole and volume fractions of each
          component of asphaltene in L phase and S phase
          respectively:

$$X_{Ai}^L = \frac{X_{Ai}^{Lmix}}{\left[\left(\dfrac{V_{Lmix}-V_A}{V_L-V_A}\right) + \left(\dfrac{V_L-V_{Lmix}}{V_L-V_A}\right)K_{Ai}\right]} \qquad (28)$$

$$X_{Ai}^S = K_{Ai} X_{Ai}^L \qquad (29)$$

$$\phi_{Ai}^L = (X_{Ai}^L V_{Ai})/V_L \qquad (30)$$

$$\phi_{Ai}^S = \left(X_{Ai}^S V_{Ai}\right)/V_A \qquad (31)$$

The average molecular weights of asphaltene in solid and liquid are then calculated.

Step 8)   The weight fraction of asphaltene precipitated (W_{AD}, gm asphaltene precipitated/gm of tank or live oil) is then obtained as follows:

$$W_{AD} = \left(\frac{V_L - V_{Lmix}}{V_L - V_A} \right) \frac{\overline{M_A^S}}{\overline{M_0}} \left(\frac{X_{AT}^0}{X_{AT}^{Lmix}} \right) \qquad 0 = LO \; or \; TO \qquad (32)$$

where $\overline{M_o}$ is the average molecular weight of oil.

Model Features, Advantages and Limitations

This model has several advantages over the homogeneous model developed previously[18]. These are as follows:

1. The model is a more realistic representation of asphaltene existence in the crude oil, i.e. they exist in large range of molecular weight distributions. Thus the limitation of using one average molecular weight to represent all asphaltene components within the crude oil is eliminated.

2. The model allows for effect of solvent molecular weight through the use of an interaction parameter between asphaltenes and asphaltene-free components.

3. The model allows for incorporation of effect of pressure and temperature on the asphaltene properties.

4. The model allows calculation of molar distribution of various asphaltene components in equilibrium solid(S), liquid (L), oil-solvent mixture (L_{mix}) and oil phases.

5. The model allows calculation of solid-liquid equilibrium constants for each asphaltene component, which provide further insight into asphaltene deposition process.

    The only limitation of the model is that it requires experimental data on asphaltene precipitation for tank oil-solvent mixtures to obtain model parameters which are needed for predictions at other pressure and temperature conditions.

APPLICATION OF MODEL TO WEST SAK OIL-SOLVENT MIXTURES

Kamath et. al.[18] reported asphaltene precipitation experimental results for various mixtures of West Sak tank oil and solvents such as ethane, carbon dioxide, propane, n-butane, n-pentane, n-heptane, a natural gas liquid (NGL) and Prudhoe Bay natural gas (PBG) at various solvent to oil ratios. These data were used to fit the model parameters of this model. Table I shows a comparison of experimental amounts of asphaltene precipitated and predicted values. The average error between the experimental and predicted values is about ± 5.3%. Table II shows the optimized model parameters. Figure 3 through 6 show the molar distribution (moles) of various asphaltene components as a function of their molecular weight in tank oil, equilibrium liquid phase and equilibrium solid phase for West Sak tank oil-n-pentane mixtures at various solvent to oil ratios varying from 2 to 20.2. These results show that as more n-pentane is added to tank oil the Wt% asphaltene precipitation increases and then levels off. Also, the heavier asphaltenes (higher molecular weight) tend to precipitate first, and remain in solid phase whereas the lower molecular weight asphaltenes tend to dissolve and remain in equilibrium liquid phase. This means that the solid-liquid equilibrium constant increases with increase in molecular weight of asphaltenes.

The model was also used to predict the amount of asphaltene precipitation for CO_2-West Sak oil mixtures at pressures up to 4000 psia. The comparison of the predicted asphaltene precipitation amounts with the experimentally measured values is given in Table III. The results in Table III show that both the experimental results and model predictions indicate considerable lower amounts of asphaltene precipitation at higher pressures than for CO_2-tank oil mixtures.

CONCLUSIONS

1. A generalized multicomponent molecular thermodynamic model has been developed to represent asphaltene equilibria and to calculate solid-liquid equilibrium constants of asphaltene components, molar distribution of asphaltenes in equilibrium solid and liquid phases and to predict the amount of asphaltene precipitation from reservoir oil upon influence of a injected solvent for different pressure, temperature and compositions.

2. The heterogeneous models such as presented here are more realistic and provide insight into asphaltene precipitation phenomena than the homogeneous type models which treat asphaltenes by a single-pseudo component.

Table I. Comparison of Experimental and Predicted Amounts of Asphaltene Precipitation for Various Solvents - West Sak Tank Oil Mixtures

Solvent Used	Solvent/Oil Ratio by Weight	Wt% Asph. (Experimental)	Wt% Asph. (Predicted)
Ethane	3.0	5.14	5.19
	13.0	5.28	5.22
CO_2	3.6	5.56	5.73
	13.6	5.80	5.75
	15.3	5.89	5.76
Propane	4.7	8.87	8.88
	7.8	8.60	8.88
	16.3	9.07	8.89
N-Butane	2.0	6.22	6.05
	8.1	6.36	6.55
	13.3	6.45	6.55
N-Pentane	2.0	1.79	1.93
	7.0	6.20	6.25
	12.5	6.30	6.20
	20.2	6.69	6.20
N-Heptane	2.0	3.83	3.32
	7.3	5.34	6.56
	13.5	7.50	6.56
	20.1	7.32	6.56
NGL	3.2	5.73	5.66
	7.6	8.80	8.76
	9.9	8.88	8.78
PBG	1.59	5.49	4.32
	1.70	7.40	6.26
	5.85	8.15	8.12
	10.42	8.08	8.12

Table II. Optimized Model Parameters for West Sak Oil - Solvent
Mixtures

SOLVENT	β	σ	a'	b'
Ethane	0.01979	166.0	0.09995	-1.495E-07
Carbon Dioxide	0.02187	166.0	0.09975	-1.7066E-07
Propane	0.03387	166.0	0.10966	-1.5414E-07
PBG	0.031	166.1	0.61094	-0.006555
N-Butane	0.025	166.0	0.60979	-0.005767
N-Pentane	0.02365	166.0	0.60896	-0.005776
N-Heptane	0.025018	166.0	0.60963	-0.0044638
NGL	0.03344	166.0	0.60975	-0.0058807

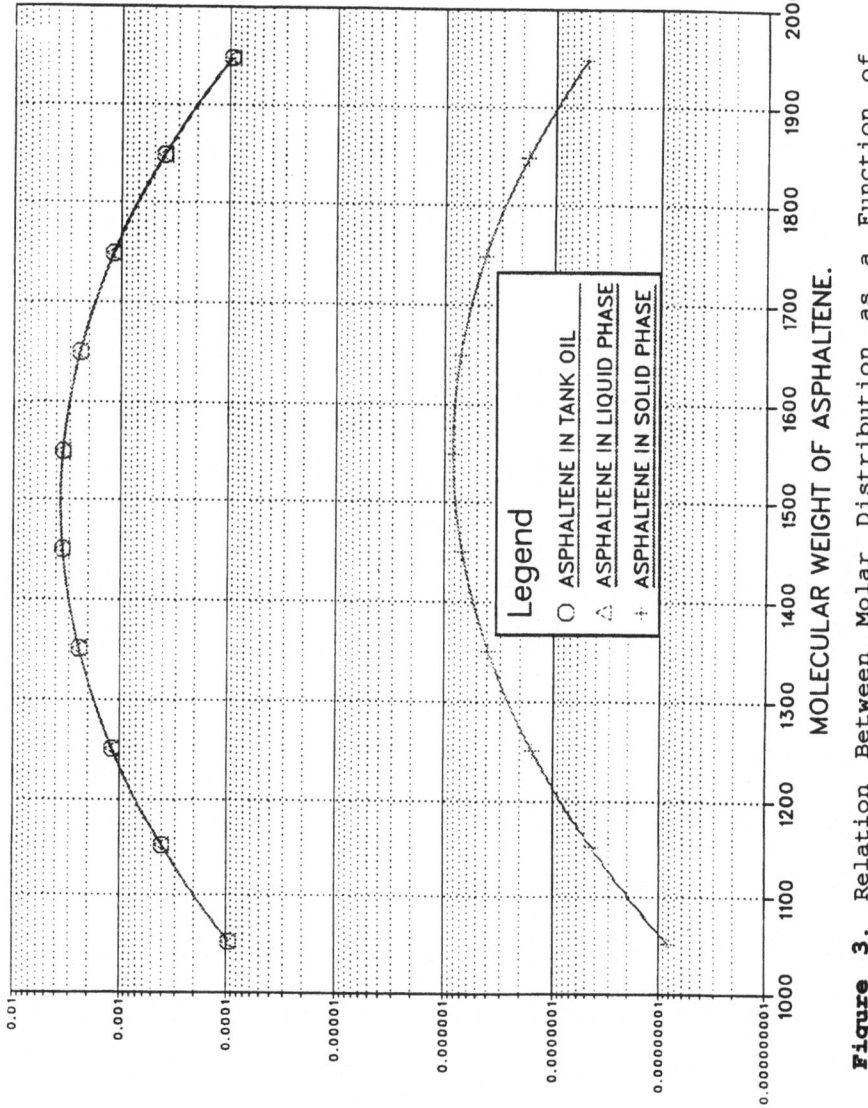

Figure 3. Relation Between Molar Distribution as a Function of Molecular Weight of Asphaltene for N-Pentane-West Sak Tank Oil Ratio = 2.0

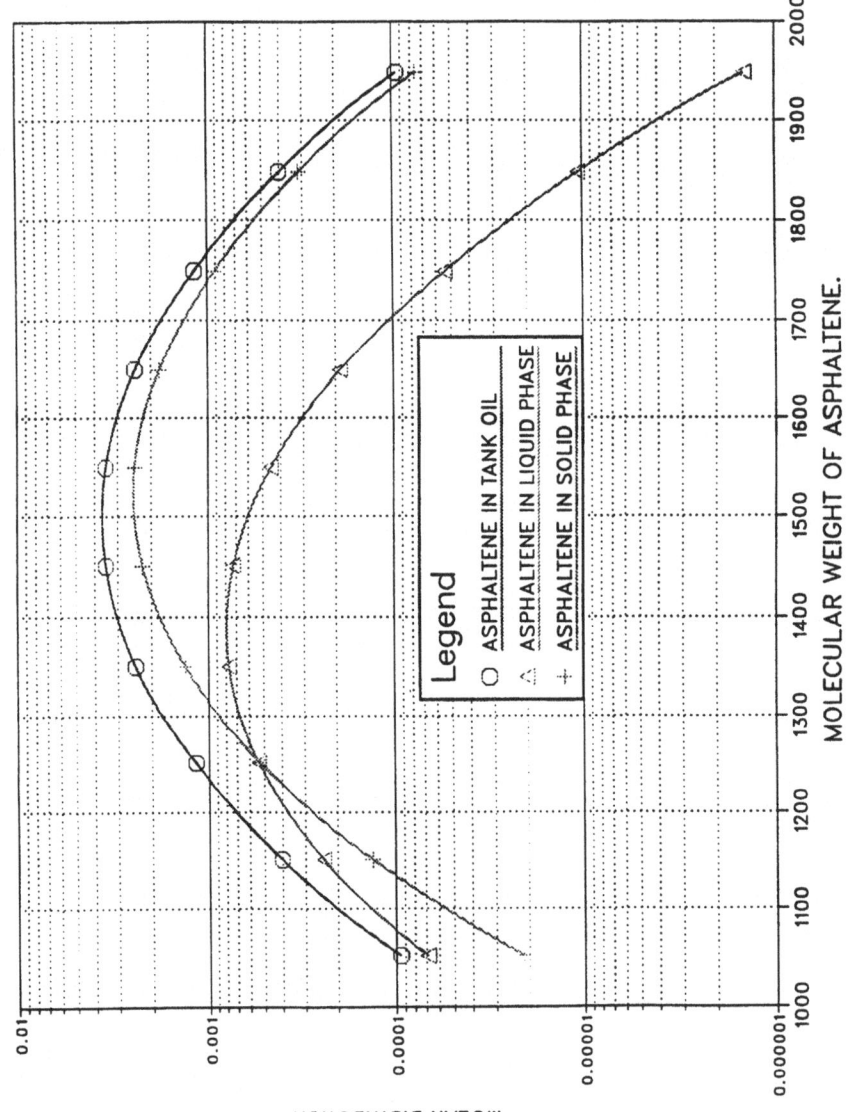

Figure 4. Relation Between Molar Distribution as a Function of Molecular Weight of Asphaltene for N-Pentane-West Sak Tank Oil Ratio = 7.0

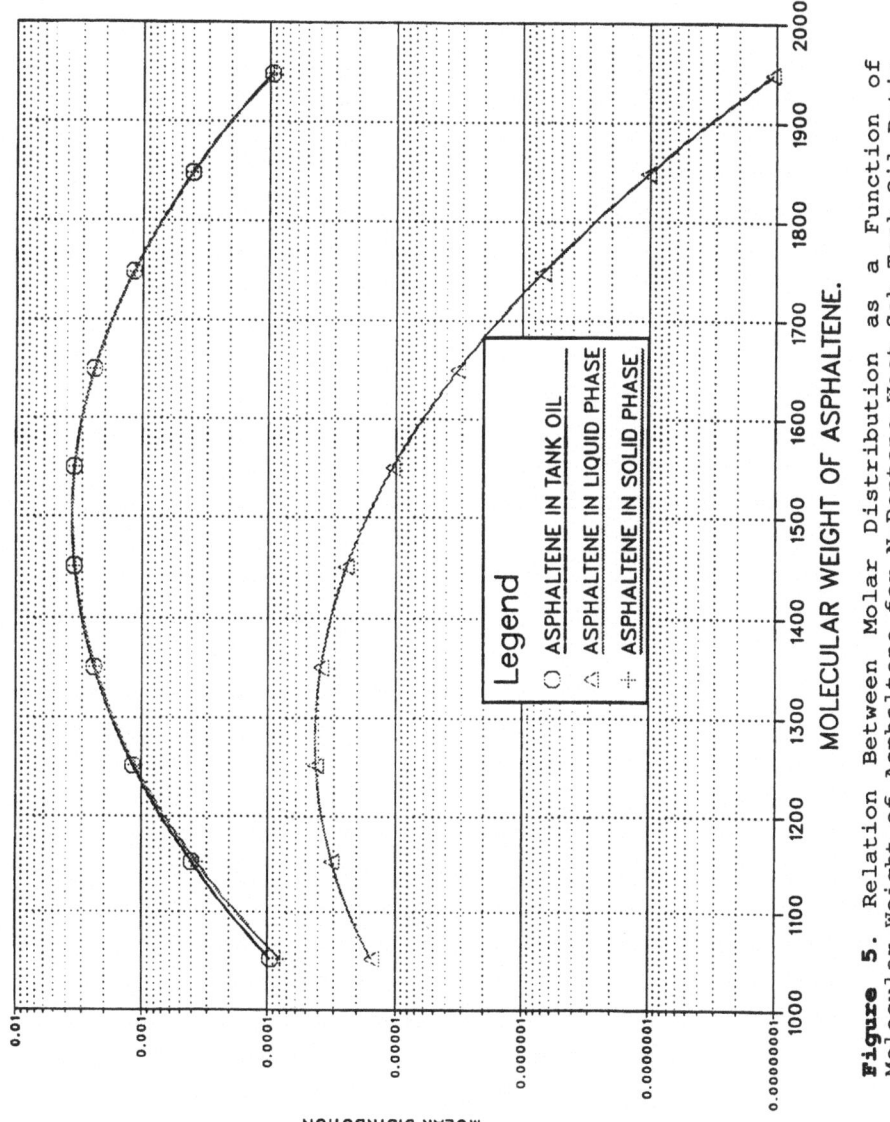

Figure 5. Relation Between Molar Distribution as a Function of Molecular Weight of Asphaltene for N-Pentane-West Sak Tank Oil Ratio = 12.5

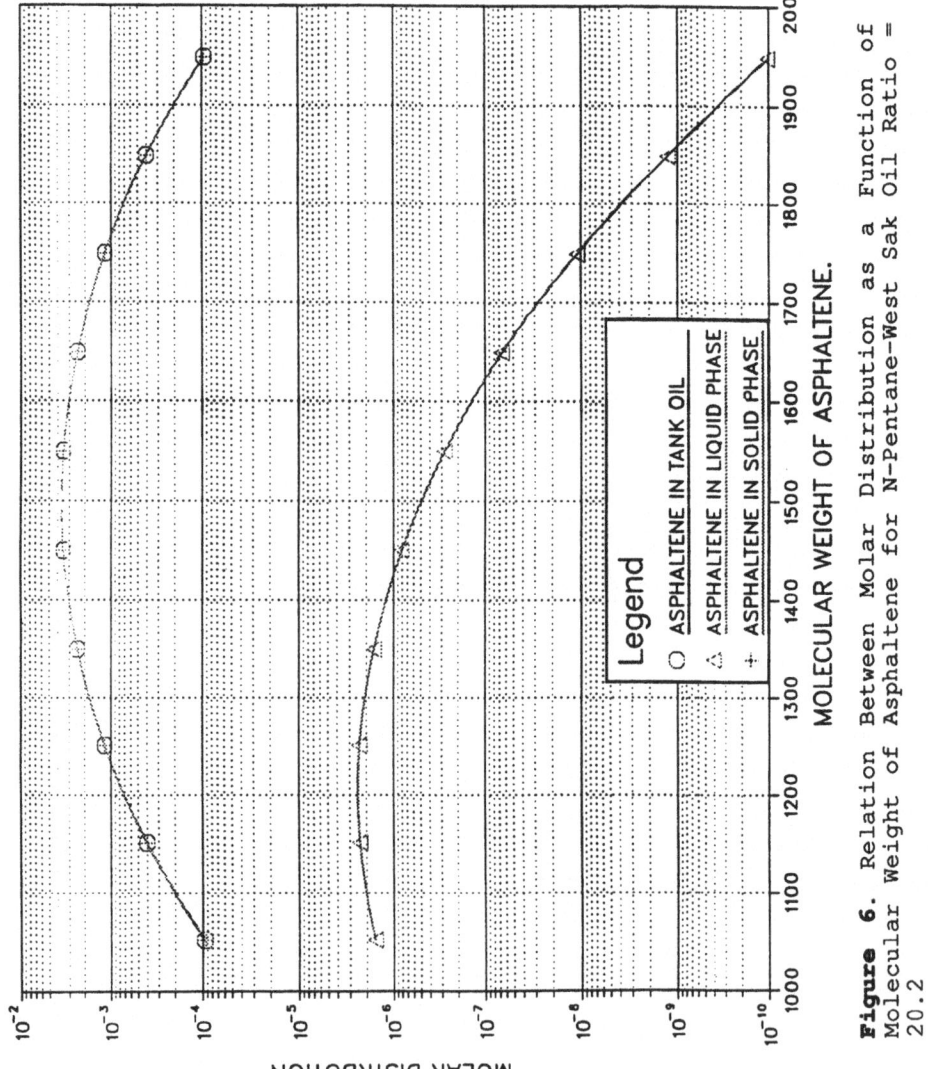

Figure 6. Relation Between Molar Distribution as a Function of Molecular Weight of Asphaltene for N-Pentane-West Sak Oil Ratio = 20.2

3. The model results show that the solid-liquid
   equilibrium constants increase with molecular weight of
   asphaltenes and the higher molecular weight asphaltenes
   precipitate out first where as the lower molecular
   weight asphaltenes tend to remain dissolved in liquid
   phase.

Table III. Comparison of Experimental Results and Model
Precipitation for CO_2 - West Sak Oil Mixtures At Higher Pressures

Solvent	Solubility Parameter (δ)	Eluted Fraction	Solubility Parameter (δ)
Hexane	7.3	Saturates	< 6
Toluene	8.9	Aromatics	< 6
CH_2Cl_2: CH_3OH 95 : 5	10.5	Resins	6 - 7
		Asphaltenes	7 - 9.2

ACKNOWLEDGEMENT

This research is supported by the U.S. Department of
Energy under the Grant #DE-FG07-90ID12839 and this financial
support is appreciated.

REFERENCES

1. Danesh, A., Krinis, D., Henderson, G.D., and Peden,
   J.M.: "Asphaltene Deposition in Miscible Gas Flooding
   of Oil Reservoirs" Chem. Eng. Res. Des. Vol. 66
   (July 1988) pp. 339-344.

2. Monger, T.G., and Trujillo, D.E.: "Organic Deposition
   During Carbon Dioxide and Rich-Gas Flooding" SPE Paper
   #18063, Proceedings of SPE Ann. Tech. Conf. &
   Exhibition, EOR/Gen. Pet. Eng. Vol., Houston, TX
   (Oct. 2-5, 1988) pp. 63-74.

3. Kokal, S., Najman, J., Sayegh, S.G., and George, A.: "Asphaltene Precipitation During Enhanced Recovery of Heavy Oils by Gas injection", CIM/AOSTRA Paper 91-10 <u>Presented at CIM/AOSTRA Tech. Conf., Banff; Canada</u> (April 21-24, 1991).

4. Hansen, P.W.: "A CO_2 Tertiary Recovery Pilot-Little Creek Mississippi". SPE Paper 6747, <u>Presented at the SPE Annual Fall Tech. Conf. & Exhibition, Denver, CO</u> (Oct. 9-12, 1977).

5. Stalkup, F.I.: "Carbon Dioxide Miscible Flooding, Past, Present and the Outlook for Future" <u>J. Pet. Tech.</u> (Aug. 1978) pp. 1102-12.

6. Harvey, M.T., Shelton, L.J., and Kelm, C.H.: "Field Injectivity Experiences with Miscible Recovery Projects Using Alternate Rich Gas and Water Injection", <u>J. Pet. Tech.</u> (Sept. 1977) pp. 1051-55.

7. Mansoori, G. A.: "Asphaltene Deposition: An Economic Challenge in Heavy Petroleum Crude Utilization and Processing", <u>OPEC Review</u> (Spring 1988) pp. 103-113.

8. Jacobs, I.C., and Thorne, M.A.: "Asphaltene Precipitation During Acid Stimulation Treatments", SPE Paper # 14823, <u>Presented a Seventh SPE Symposium on Formation Damage Control, Lafayette, LA</u> (Feb. 26-27, 1986).

9. Jacobs, I.C.: "Chemical Systems for the Control of Asphaltene Sludge During Oilwell Acidizing Treatments", SPE paper #18475, <u>Presented at SPE Int. Symp. on Oilfield Chem. Houston, TX</u> (Feb. 8-10, 1989).

10. Leontaritis, K.J.: "Asphaltene Deposition: A Comprehensive Description of Problem Manifestations and Modeling Approaches" SPE Paper #18892, <u>Presented at SPE Prod. Op. Symp. Oklahoma City, OK</u> (March 3-4, 1989).

11. Collins, S.H., and Melrose, J.C.' "Adsorption of Asphaltenes and Water on Reservoir Rock Minerals", SPE Paper # 11800, <u>Presented at Int. Symp. on Oilfield and Geo. Chemistry, Denver, CO</u> (June 1-3, 1983).

12. Dubey, S.T., and Waxman, M.H.: "Asphaltene Adsorption and Description From Mineral Surfaces" SPE Paper # 18462 <u>Presented at SPE Int. Sym. on Oilfield Chemistry, Houston, TX</u> (Feb 8-10, 1989).

13. Kim, S.T., Boudh-Hir, M.D., and Mansoori, G.A.: "The Role of Asphaltene in Wettability Reversal" SPE #20700, <u>Presented at 65th SPE Ann. Tech. Conf. and Exh. New Orleans, LA</u> (Sept 23-28, 1990).

14. Yang, J.: "Effect of Asphaltene Deposition on Rock-Fluid Properties" M.S. Thesis, U. Alaska Fairbanks (Aug. 1992) pp. 1-104.

15. Fussel, L.T.: " A Technique of Calculating Multiphase Equilibria", *SPEJ.* (Aug. 1979) pp. 203-209.

16. Hirschberg, A., deJong, L.N.J., Schipper, B.A., and Meijer, J.G.: "Influence of Temperature and Pressure on Asphaltene Flocculation" SPE Paper # 11202, *Presented at SPE Ann. Tech. Conf. & Exh., New Orleans, LA* (Sept. 26-29, 1984).

17. Burke, N.D. Hobbs, R.D., and Kashou, S.F.: "Measurement and Modeling of Asphaltene Precipitation from Live Reservoir Fluid Systems" SPE 18273, *Presented at 63rd Ann. Tech. Conf. & Exh. of SPE, Houston, TX* (Oct. 2-5, 1988).

18. Kamath, V.A., Islam, M.R., Patil, S.L., Jiang, J.C., and Kakade, M.G.: "The Role of Asphaltene Aggregation in Viscosity Variation of Reservoir Hydrocarbons and in Miscible Processes", *Particle Tech. & Surface Phen. in Min. & Pet.* Plenum Publ. Corp. (1991) pp. 1-22.

19. Novasad, Z., and Costain T.G.: "Experimental and Modeling Studies of Asphaltene Equilibria for a Reservoir Under CO_2 Injection" SPE Paper #20530 *Presented at the 65th SPE Ann. Tech. Conf. & Exh., New Orleans, LA* (Sept 23-26, 1990).

20. Kawanaka, S., Park, S.J., and Mansoori, G.A.: "The Role of Asphaltene Deposition in EOR Gas Flooding: A Predictive Technique", SPE/DOE Paper #17376, *Presented at the SPE/DOE EOR Symp; Tulsa, OK* (April 17-20, 1988).

21. Leontaritis, K.J., and Mansoori, G.A.: "Asphaltene Flocculation During Oil Recovery and Processing. A Thermodynamic Colloidal Model' SPE paper # 16258, *Presented at the SPE Int. Symp. Oilfield Chem.; San Antonio, TX* (Feb. 4-6, 1987).

22. Twu, C.H.: "An Internally Consistent Correlation for Predicting Critical Properties and Molecular Weights of Petroleum and Coal-Tar Liquids", *Fluid Phase Equil.* Vol. 16, (1984), pp. 137-150.

23. Gambill, W.R.: "Determine Heat of vaporization", *Chem. Eng.*, Vol. 64, No. 12, (Dec. 1957), pp. 261-266.

24. Scott, R.L., and Magat, M.: "The Thermodynamics of High Polymer Solutions: I. The Free Energy of Mixing of Solvents and Polymers of Heterogeneous Distribution" *J. Chem. Physics*. 13(5), (May 1945), pp. 172-177.

FRACTIONATION OF ASPHALT BY THIN-LAYER CHROMATOGRAPHY INTERFACED WITH FLAME IONIZATION DETECTOR (TLC-FID) AND SUBSEQUENT CHARACTERIZATION BY FTIR

H. J. Lian, C. Lee and T. F. Yen

Environmental and Civil Engineering
University of Southern California
Los Angeles, CA 90089-2531

ABSTRACT

A simple, effective, and inexpensive thin-layer chromatography interfaced with flame ionization detector (TLC-FID) method is a major tool for quantitative analysiss of heavy hydrocarbon compositions, especially for asphalt composition. This destructive method is not only faster than ASTM D4124 or SARA, but also quicker than preparative Chromatotron since there is no need to spend time for evaporation of solvent. The results will be complemented by Fourier transform infrared spectrometer (FTIR) for special identification of asphalt. This method can be applied in practice for separation of chemical substances especially in medicine and biology, lipid chemistry and the characterization of crude oil.

INTRODUCTION

Asphalt consist of four fractions: saturates, aromatics, resins and asphaltenes which can not be separated by gas chromatography (GC) owing to high molecular weight and low volatility. It is also not practical by using high performance liquid chromatography (HPLC) due to high cost of packing columns and low efficiency by irreversible adsorption of compounds on column.[1-3]

In general, asphalt can be separated by column chromatography (ASTM D4124[4], SARA[5], solvent fractionation[6]), or thin layer chromatography (preparative Chromatotron[7], destructive TLC-FID[8]). In this project we attempt to use thin-layer chromatography interfaced with flame ionization detector (TLC-FID) for asphalt separation because it uses less solvent (about 600 milliliter instead of ASTM D4124 and SARA which use 4-6 liter for one experiment) and it consumes only 3 hours per run for TLC-FID instead of ASTM D4124, SARA which consumes about 2-3 days per run. This destructive TLC-FID method still has one advantage over preparative Chromatotron which is time saving; because sample fractions collected in each beaker by preparative Chromatotron still need to wait for of hours until solvents evaporate away. Destructive TLC-FID does not need this.

Padley[9] and Szakasits[10] started to use TLC-FID in 1969-1970. Later, many researchers applied this method in biology, medicine, pharmacy, food industry, chemical industry quite successfully[11-14]. Selucky was the first person to apply TLC-FID on coal-derived liquid and the result was very compatible with ASTM D4124 and SARA if correction factors were figured out carefully[15]. Poirier[16] and Yoshida[17] used the same method but with different rods and solvents to identify coal derived liquids and the results were quite similar. Here, we attempt to use TLC-FID with SIII rods in three solvents: hexane, toluene, dichloromethane/methanol (at 95/5) ratio to separate asphalt into four fractions. The details will be described in the following section. This tool also can be used to identify asphalt components which were sonicated by ultrasound and the results will be supported by Fourier transform infrared spectroscopy (FTIR).

EXPERIMENTAL

The Iatroscan TH-10, Mark III Analyzer and SIII Chromarods were used for asphalt separation, and seven samples of asphalt, which were supplied by Strategic Highway Research Program (SHRP), were used for these experiments. The source and refining process of seven different asphalts are listed on Table I. Also, three solvents were selected to elute different fraction of asphalt, and the detailed procedure are listed on Figure 1.

Initially, ten Chromarods were cleaned and activated by hydrogen flame, then seven sample solutions each with 500 mg of the seven asphalts (Table I) were dissolved in 50 ml of cyclohexane and heated in an oven at 60 °C. Seven sample solutions with 10 $\mu g/\mu l$ concentration were prepared. One μl of each sample solution was applied on Chromarods by a 1 μl microsyringe.

Table I. Source and refinery process of the samples

Sample	Source	Refinery Process
AAA-1	Lloydminister	distillation
AAB-1	Wyoming Sour	distillation
AAD-1	California Coast	distillation
AAE-1	Lloydminister	air-blown
AAG-1	California Valley	distillation
AAK-1	Boscan	distillation
AAM-1	West Texas	solvent

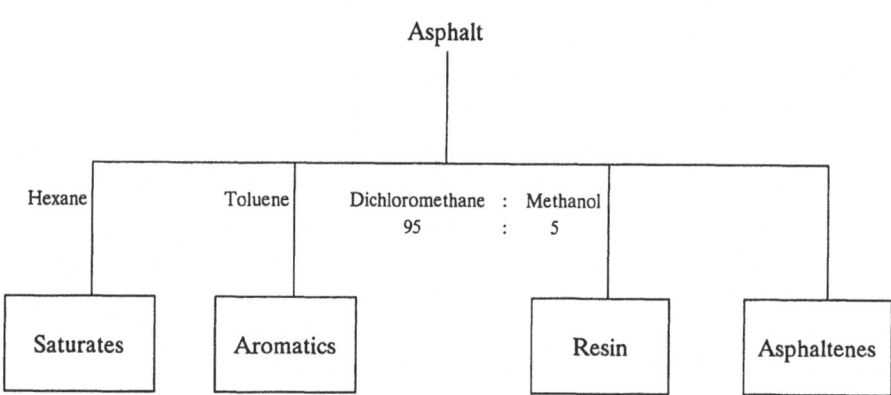

Figure 1. Characterization of core asphalt by TLC-FID

Table II. The separation of seven core asphalts by TLC-FID

Compound	AAA-1	AAB-1	AAD-1	AAE-1	AAG-1	AAK-1	AAM-1
Saturates (%)	13.0	11.7	12.3	11.4	9.3	7.0	9.2
Aromatics (%)	38.9	32.8	23.5	15.8	36.7	31.6	35.7
Resins %	30.7	36.9	40.8	42.6	49.1	37.7	41.0
Asphaltenes (%)	17.4	18.6	23.4	30.2	4.9	23.7	14.1

Three mobile phases, n-hexane, toluene, dichloromethane:methanol, 95:5 (V:V), were used stepwise to elute saturates at 10 cm, aromatics at 5 cm, and resins at 2.5 cm of rod location which was scaled on rod holder, and the most polar asphaltenes were remained at the original (0 cm position) place. During each step Chromarods should be placed for 2 minutes in the hood under a hot lamp in order to remove the solvent after eluting soluble components to the suitable position, and then placing them into a constant humidity chamber with 65% relative humidity retained by 35.8% sulfuric acid solution for 10 keep to make the degree of water adsorption constant[18].

Finally, Chromarods were scanned over the full surface area by FID. The individual separated zones were ionized in a hydrogen flame and the ionization current produced was amplified to feed into a HP 3390A integrator. The conditions of scanning were scan speed - 30 sec/rod, gas flow of hydrogen, 0.9 kg/cm^2 of air, 2000 ml/min; chart speed, 2 cm/min. Results of seven core asphalts are listed in Table II.

The second experiment was using AAG-1 asphalt dissolved in chloroform, then oxidized by hydrogen peroxide under ultrasound. With the same technique, it is easy to distinguish the change of four fractions after sonication, and the results can be supported by FTIR. First, 5 g of AAG-1 asphalt were dissolved in 50 ml of chloroform to make a 10 percent sample solution. Then the sample solution was stirred at 150 rpm by a magnetic stirring bar, and gradually oxidized for one hour by adding 2% hydrogen peroxide under ultrasound. A VC 50 microprocessor was used for energy supply of ultrasound. After that, sample solution was diluted to 2% using the same procedure, which was mentioned above, to identify four fractions of sonicated AAG-1 asphalt. Figure 2 shows the four fractions of AAG-1 asphalt at different reaction time.

Ten milliliter of sonicated 2% AAG-1 solution were added to 100 milliliter of n-pentane to precipitate asphaltene. Then asphaltene was filtered and dried completely in vacuum oven at room temperature. After that, 0.002 gm of sonicated AAG-1 asphalt were mixed with 0.2 gm of potassium bromide to make a 1.5 centimeter diameter of pellet. Finally, the pellet was mounted on a sample holder, and scanned by Nicolet 5DX Infrared Spectrometer for 200 times. Figure 3 presents the IR spectrum of AAG-1. Two areas A1 and A2 are marked to correlate the asphaltene content. Fig. 4 is the plot of A_1 to A_2 ratio as a function of time.

RESULTS AND DISCUSSIONS

Table III list the solubility parameters of four fractions of asphalt[19] and three solvents[20]. The selection of the solubility parameters each solvent at a level higher than for each eluting fraction (Figure 1) is due to the

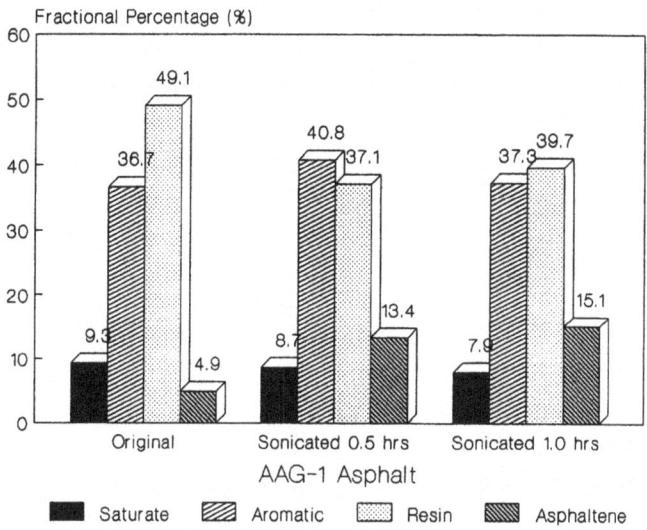

Figure 2. The separation of sonicated AAG-1 asphalt with H_2O_2 by TLC-FID method

Table III. Solubility parameters for solvent/solvent mixture and asphalt fractions

Mol% CO_2	Pressure (Psia)	Wt% Asph. (Experimental)	Wt% Asph. (Predicted)
30	900	0.177	0.124
	1300	0.183	0.097
	1705	0.197	0.195
	2500	0.151	0.067
	3000	0.179	0.079
	3500	0.183	0.102
	4000	0.212	0.151
50	1000	0.140	0.149
	1705	0.192	0.062
	3000	0.156	0.156
	4000	0.108	0.206

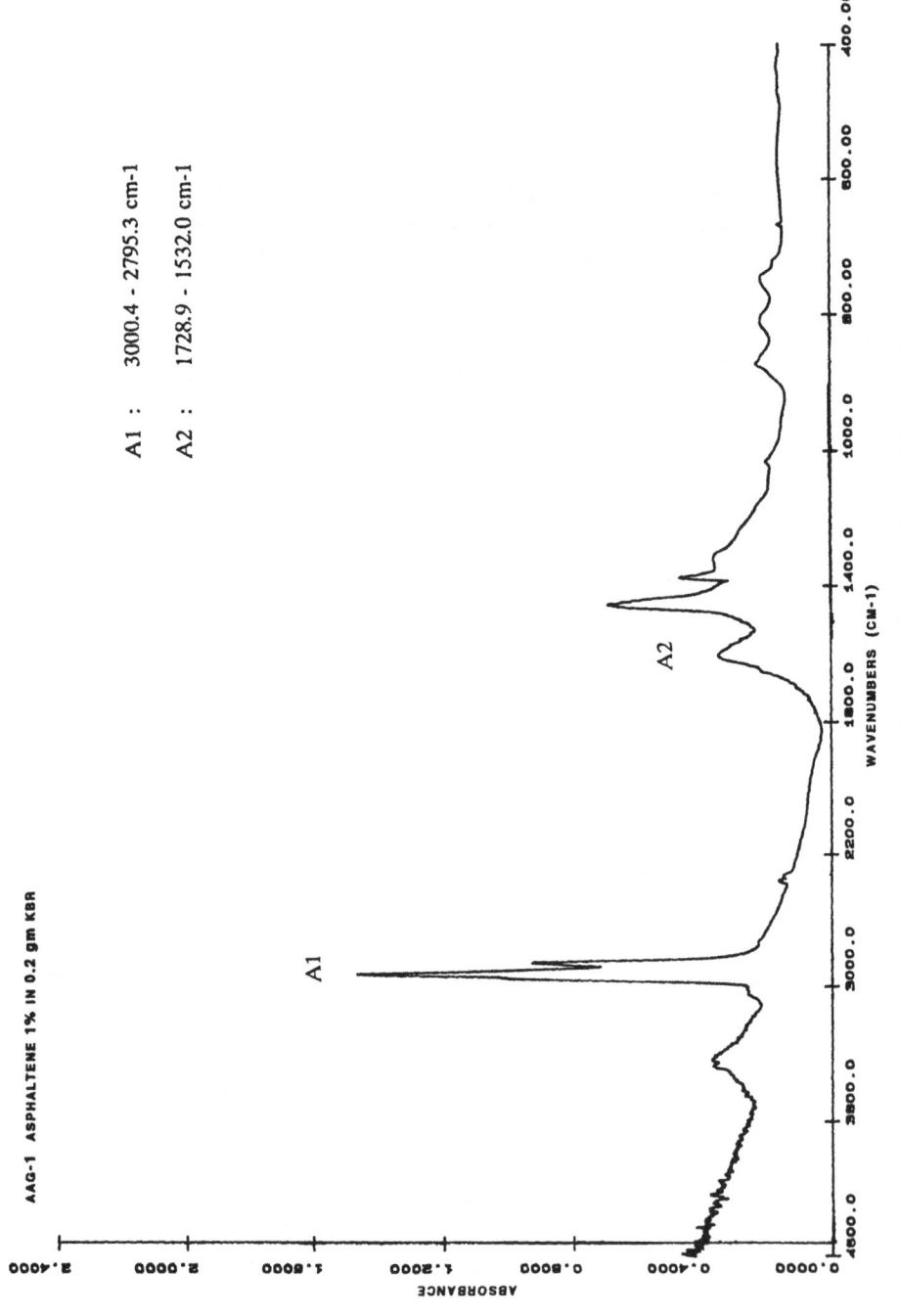

Figure 3. AAG-1 asphaltene (1% in KBr) IR spectrum

234

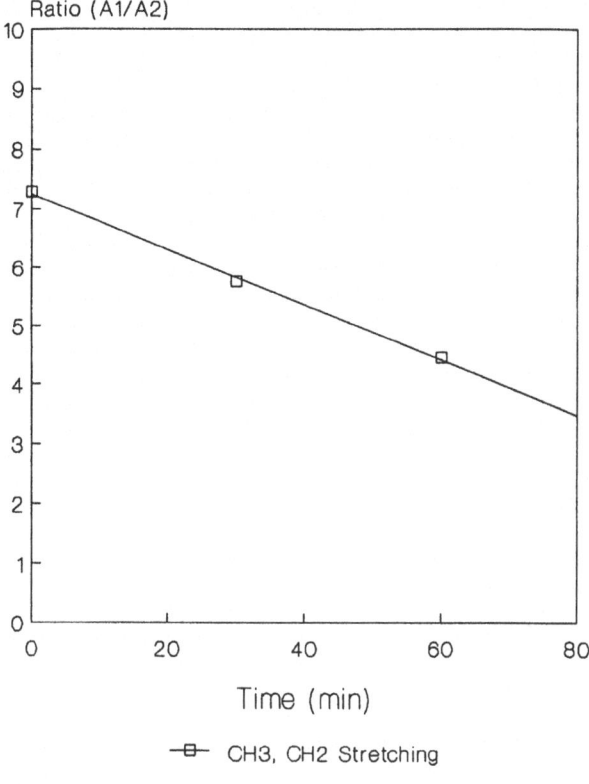

A1 : Area of 3000.4 - 2795.3 cm-1
A2 : Area of 1728.9 - 1532.0 cm-1

Figure 4. The comparison of sonicated AAG-1 asphaltene by IR spectra

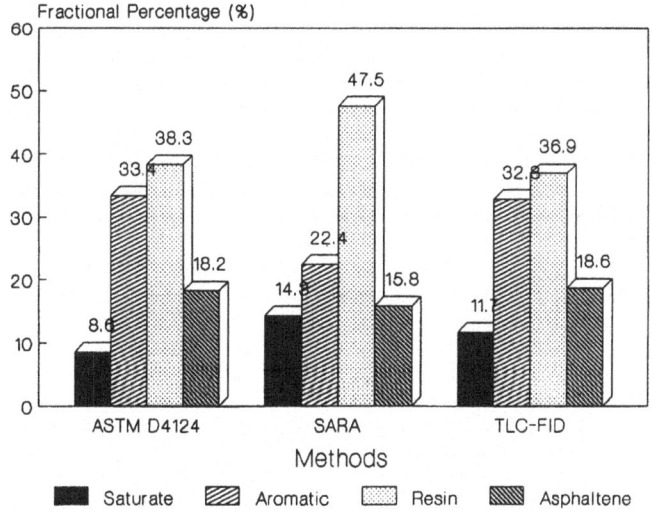

Figure 5. The comparison of AAB-1 asphalt fractions by different methods

absorption strength of stationary phase (silica gel). The results obtained in Table II were already corrected by calibration curves of each function in which standards were isolated by preparative Chromatotron[7]. Figure 5 shows that the separation of AAB-1 asphalt by TLC-FID is close to ASTM D4124 method. These data were supported by SHRP, even better than SARA method[21]. Figure 2 illustrates that asphaltene content can be increased up to 10.2% within one hour by sonication with 0.2% hydrogen peroxide. It is also proven by comparing the peak area A2 of functional group C=C, C=O with the peak area A1 of CH_2, CH_3 stretching. The result shows the decrease of A1 to A2 ratio with time (on Figure 4) is due to the increase functional group in asphaltene. Again, it demonstrates that TLC-FID is a good tool for asphalt separation.

CONCLUSIONS

TLC-FID is a fast, effective, inexpensive tool for asphalt separation. It can not only be used on the separation of heavy hydrocarbon, high molecular weight compounds, but also be used to characterization of crude oil, or be applied in biology, pharmacy, food industry, chemical industry.

REFERENCE

1. Ranny, M.; and Pokorny, J.; Comparison of TLC-FID and HPLC for the determination of oxidized products in ethyl linoleate, J. Planar Chromatography - Mod. TLC, 1(3), 255-7 (1988).

2. Selucky, M. L.; Quantitative analysis of coal-derived liquids by thin-layer chromatography with flame ionization detection, Anal. Chem., 55, 141-43 (1988).

3. Poirier, M. A. and George, A. E.; A rapid method for hydrocarbon-type analysis of heavy oils and synthetic fuels by pyrolysis thin layer chromatography, Energy Sources, (7), 151-64 (1983).

4. ASTM D4124: Standard Test Methods for Separation of Asphalt into Four Fractions, American Society for Testing and Materials, Philadelphia, PA, (1988).

5. Altgelt, K. H., Jewell, D. M., Latham, D. R. and Selucky, M. L.; Chromatography and Petroleum Analysis, Chromatographic Science Series; Marcel Dekker, New York, pp.193-196 (1976).

6. Schwager, I. and Yen, T. F.; Coal Liquefaction Products from Major Demonstration Processes I. Separation and Analysis, Fuel (London), 57, 100-104 (1978).

7. Lian, H. J., Lee, Z. H., Wang, Y. Y. and Yen, T. F.; Characterization of Asphalt with the Preparative Chromatotron, Journal of Planar Chromatography, **5**, 263-266 (1992).

8. Ranny, M.; Thin-Layer Chromatography with Flame Ionization Detection, D. Reidel Publishing Company, Boston (1987).

9. Padley, F. B.; The Use of a Flame-Ionization Detector to Detect Components Separated by Thin-Layer Chromatography, J. Chromatography, **39(1)**, 37-46 (1969).

10. Szakasits, J. J., Peurifoy, P. V. and Woods, L. A.; Quantitative Thin-Layer Chromatography Using a Flame Ionization Detector, Anal. Chem., **42(30)**, p.351-354 (1970).

11. Itoh, T., Tanaka, M. and Kaneko, H.; "Flame Ionization Detection System for TLC of Lipids", Thin Layer Chromatography(Touchstone, S.C., and Rogers, D., ed.) Wiley, New York, p.536 (1981).

12. Okumura, T., Kadano, T. and Iso'o, A.; Sintered Thin-Layer Chromatography with Flame Ionization Detector Scanning, J. Chromatography, **108(2)**, p.329-336 (1975).

13. Takase, Y., and Yoshioka, S.; Separation of Sulphanilic Acid and Sulphonamides, Iatron Labs Inc., Tokyo (1977).

14. Andree, H., Muller, C. W. R. and Schmid, R. D.; Lipases as Detergent Components, J. Appl. Biochem., **2(3)**, 218-229 (1980).

15. Selucky, M. L.; Quantitative Class Separation of Coal Liquids Using Thin-Layer Chromatography with Flame Ionization Detection, Lipids, **20(8)**:546-551 (1985).

16. Poirier, M. A., Rahimi, P. and Ahmed, S.M.; Quantitative Analysis of Coal-Derived Liquids Residues by TLC with Flame Ionization Detection, J. Chromatographic Science, **22**, 116-119 (1984).

17. Yoshida, R., Miyazawa, M., Yoshida, T., Ishizaki, K., Shinriki, N. and Maekawa, Y.; Upgrading of Coal-Derived Liquids:3. Characterization of Upgraded Liquids by Thin-Layer Chromatography combined with Flame-Ionization Detection, Fuel **65**, 421-424 (1986).

18. Iatroscan TH-10 Instrument Application : Optimizing Reproducibility with the Iatroscan TLC/FID analysis, ANCAL Inc., p.5 (1982).

19. Yen, T. F.; "Asphaltic Materials", in Encyclopedia of Polymer Science and Engineering, (M. Grayson and J. I. Krochwitz, eds.), Wiley, New York (1988).

20. Weast, R.C.; Handbook of Chemistry and Physics, 66th edition, CRC Press Inc., p. c676-678 (1985).

21. Wang, Y. Y. and Yen, T. F.;. "Strategic Highway Research Program Contract A002B Quarterly Report Task 2-Colloidal Chemical Approach to Age Hardening", p.5, March-June, (1990).

AUTHOR INDEX

SUBJECT INDEX

Coal (cont'd)
    concentration, 48, 144-146
    derived, 91, 96-98, 141
    liquefacation, 48, 91-93, 141
                143, 144, 151
    particles, 48, 141, 142, 145
                151
    structure, 48
Coefficients
    diffusion, 35, 38, 106, 109
    longitudinal, 33
    molecular diffusion, 38, 41
    transeverse, 33
Cohesiveness, 63
Coke formation, 48
Colloidal, 65, 66, 104, 115,
                145, 157, 159, 162
                167, 168, 171, 174
                176, 187, 190, 191
                194, 197, 198, 208
Composition
    reservoir fluid, 2
Co-processing
    catalytic, 48
    coal/bitumen, 61
    molten, 51
    performance, 48
    reactions, 49
    thermal, 48
Corrosion, 63
Critical temperature, 28
Crude oil, 82
Cyclics, 13, 14, 17, 27
    composition, 15

Darcy's Law, 34, 35
Demetallation, 48
Deoxygenation, 48
Desorption, 117, 120, 121
Desulphurization, 48
Dielectric constant, 3
Diffusion, 27, 109, 120, 130,
                167-169
Diffusivity, 108, 167-169
Dispersion, 33
Dissociation, 96, 116, 117
Dissolution, 33, 115, 130
Distillation, 48
Distribution, 33, 141, 151, 152
                156, 158-161, 163
                167, 169, 171, 176
                179, 205, 210, 211
                215, 217, 221, 222
                223, 224
Downstream, 81

Efficiency, 91
Electron micrographs, 106
Elemental, 82, 87, 115
Emulsion, 94, 195

Environmentally, 101, 102,
                108, 111
Equation, 27, 117, 145, 173,
                175, 188, 214, 215
    algebraic, 35
    diffusion-adsorption, 34
    flow, 34
    Grimson-Barker, 162
    Kozeny-Carman, 34
    Ornstein-Zernike, 159
Exterior, 14
Extraction, 14, 15, 94, 115,
                175

Feed-stock, 81, 89
Filterate, 54
Flocculate, 81, 82
Fluxes
    convective, 34
    diffusive, 34
Fouling, 87
Fractionation, 156, 229
Fractions, 63, 82, 86, 87,
                89, 93, 94, 115,
                126, 128, 142,
                149, 151, 168
Fragmentation, 26

Gravitational forces, 2
Gravity, 2

Heavy oil, 1, 10, 11, 31, 81
                93, 123, 125,126,
                130, 133, 136
    conversion, 81
    recovery, 125
    reservoirs, 32, 130, 135,
                136
Heptane, 14, 15, 26, 81, 82,
                157, 185, 186, 191
Heteroatoms, 87, 92, 115,
                168
High saturation zone, 32
Hydrocarbon, 2, 95, 101, 104
                106
Hydrocraked, 4
Hydrocraking, 6, 48, 49
Hydrogenation, 58, 91, 92
Hydrogen pressure, 47, 60,
                82, 91,
                94
Hydroliquefaction, 48
Hydro-processing, 87
Hysteresis, 186

Impermeability, 63
Injection, 35, 130
    pressure, 35, 43
    rate, 38, 39

The manufacturer's authorised representative in the EU is Springer
Nature Customer Service Centre GmbH, Europaplatz 3, 69115 Heidelberg,
Germany. If you have any concerns regarding our products, please
contact ProductSafety@springernature.com

Printed and bound by CPI Group (UK) Ltd, Croydon, CR0 4YY
29/04/2026
02099472-0016